BATTLESHIPS

BATTLESHIPS

The World's Greatest Battleships from the 16th Century to the Gulf War

DAVID ROSS

amber
BOOKS

First published in 2022

Reprinted in 2025

Published by Amber Books Ltd
United House
North Road
London N7 9DP
United Kingdom
www.amberbooks.co.uk
Instagram: amberbooksltd
Facebook: amberbooks
Twitter: @amberbooks
Pinterest: amberbooksltd

ISBN: 978-1-83886-218-3

Project Editor: Michael Spilling
Designer: Brian Rust
Picture Research: Terry Forshaw

Printed in China

PICTURE CREDITS
AKG: 41 (Oronoz)
Alamy: 24 (MEPL), 25 (MEPL), 77 (19th Era), 81 (World History Archive)
Amber Books: 142, 143, 171
Art Archive: 28
Art-Tech/Aerospace: 61, 166, 186, 187, 194, 198, 202, 219
Cody Images: 56, 57, 65, 118 bottom, 122, 123, 131, 134, 135, 151, 154, 155, 158, 170, 174, 175, 178, 183, 190, 195, 203, 207, 210, 218
Corbis: 37 & 44 (Gallery Collection), 69 (Lebrecht Music & Arts)Dreamstime: 20 (Martin Maun), 33 (Michael Wood)
Mary Evans Picture Library: 17, 101, 114
Getty Images: 52 (Hulton)
Library of Congress: 8, 88, 89, 95, 96, 97, 104, 110, 126, 127, 139
U.S. Department of Defense: 6, 7, 85 both, 106, 211, 215

All artworks ©Tony Gibbons/Amber Books, except for the following:
Art-Tech: 10/11, 12, 13, 14/15, 18/19, 22/23, 26/27, 30, 31, 34/35, 45, 46/47, 50/51, 62/63, 111, 128/129, 160/161, 180/181, 184/185, 206

Contents

Introduction

For more than 400 years, the big-gun warship stood as the supreme achievement of naval architecture and ship equipment. It was not only a major instrument of warfare, but a visible emblem of a nation's power, wealth and pride. Possession of a powerful fleet had a major influence on a country's foreign policy.

The first capital ships were intended as flagships – conspicuous by their size, flying the banner of an admiral, and acting as lead vessel to a squadron or fleet of smaller craft. By the end of the sixteenth century, their value as fighting machines encouraged the maritime countries of Europe to build more big warships, and also led to a change in the tactics of war at sea. The concept of the 'line of battle' gradually became dominant, in which two fleets formed up in line to bombard each other. Such encounters were always more than a simple slugging match: the advantage would rest with the admiral who kept the wind in his favour and could place his ships in a position to do maximum damage. A line-of-battle ship, or 'ship of the line', had to be big and broad enough to carry heavy cannon and act as a steady firing platform, and strong enough to withstand reciprocal fire. These remained the two fundamental requirements even after wooden hulls gave way to iron and steel, sails were supplanted by steam power, and cannonballs were replaced by armour-piercing or high-explosive shells. Eventually the name was shortened to 'battleship', though the term did not come into common use until the nineteenth century.

Developments in the making of armour plate and in the design of guns strongly influenced battleship construction in the nineteenth century, with progress in one field forcing new attention to the other. Many one-off ships were launched, some of them to have brief careers. For a time, the fashioning of the torpedo into an effective weapon looked as if it might bring the age of the battleship to a premature end. But by the early 1900s, innovations and improvements established the combination of internal

Commissioned in 1912, USS *Arkansas* served in both world wars, including providing fire support for the amphibious assaults of Normandy, Iwo Jima and Okinawa.

USS *Iowa* demonstrates the power of its 406mm (16in) guns. Commissioned in 1943, *Iowa* served in the US Navy until 1990.

armoured decks and bulkheads, big guns mounted on the centre-line, with super-firing turrets, and turbine propulsion, which would become standard in the twentieth century, and culminate in such mighty warships of World War II as Germany's *Bismarck*, the USA's *Iowa* and Japan's *Yamato*.

Advent of the battlecruiser

Also in the 1900s, a variant on the capital ship appeared in the form of what quickly became known as the battlecruiser, built by several navies, notably the British, German and Japanese. The US Navy built three 'large cruisers' in 1943–45, classified as CB rather than BB for battleship. In effect the battlecruiser was a battleship in which, to a degree, weight of armour was exchanged for increased speed. While some purists might argue that the battlecruiser is a separate category of warship, a few have been included in this book. Since they took part in combat with battleships, and in most cases were bigger, more expensive and often carried heavier guns, than many battleships, this seems reasonable.

In any naval strategy up to 1942, battleships were the prime elements. Yet they fought relatively few full-scale battles. At Tsushima in 1905 the Japanese battle fleet destroyed that of Imperial Russia. In 1914–18, the admiral commanding the British Grand Fleet was known as the only man who could lose the war in a day, if the Kaiser's High Seas Fleet were to wrest control of the North Sea from the Royal Navy. The two fleets met only once, in the inconclusive Battle of Jutland in May 1916. In World War II, battleships were frequently in action, though generally in escort roles or as supporting units in large-scale operations involving many other ships and aircraft.

Battleship building boom

Many hundreds of men-of-war were built up to the end of the nineteenth century. Between 1905, the year before the 'revolutionary' HMS *Dreadnought* appeared, and 1946, when HMS *Vanguard* became the last new battleship to be commissioned, some 220 capital ships were put in commission by the world's naval powers (not all of them, of course, in service at the same time). Great Britain built the largest number, with around 52, the USA had 41, Germany had 40, Japan and France each had 20, Italy had 19, Russia 13, Austria-Hungary had seven, and Argentina, Brazil and Chile each had two. Many more were planned or begun but not completed, due to wars coming to an end, or international treaty agreements, or simply lack of resources. They were hugely expensive. The battlecruiser HMS *Hood* cost £6,025,000 to build in 1916–20, and £400,000 a year to run.

This book sets out to provide, through major entries on 52 ships, a guide to the world's great capital ships. Some of them had eventful careers and participated in famous actions, others are remembered only for being sunk. So powerful was the link between the battleship and national feeling that even into the nuclear era, some, like France's *Richelieu*, were held in naval reserve. As late as 1967–68 USS *New Jersey*, of the *Iowa* class, was brought back into commission. The aircraft carrier (its first examples built on hulls intended for battlecruisers, such as HMS *Courageous*, featured in this book) and the nuclear submarine have since replaced them, but their place in the history of ships and navies is assured.

Wooden Walls to Armour Plate

Between the first men-of-war capable of firing a broadside and the iron-hulled, steam-driven battleships with gun turrets mounted on the centre-line, lie four centuries of experiment, advance and rivalry. While navies of many nations competed to build ships that would outfight any potential enemy, the two essentials always remained the same: firepower and staying power.

Opposite: A first-class capital ship completed in 1895, USS *Indiana* was classified as a coastal battleship for domestic political reasons.

Henry Grace à Dieu (1514)

Known as 'Great Harry', this was one of the first English ships to be designed specifically for warfare. Very large for its time, it reflected the aspirations of the English state during a period of intense rivalry and occasional warfare with France.

In the reign of Henry VIII English policy towards Europe began to take the form it would maintain for the next five centuries, to try to establish a balance of power on the continent so that no one nation became so strong as to be able to defeat the others, and threaten England. France and French power were the particular source of concern at that time. Sea power was necessary to enforce English policy and also to display the prestige of the English state and the grandeur of its monarch. Purpose-built warships were, literally, the King's ships, and Henry took considerable pride in his three large ships, the other two being *Mary Rose* (1511) and the oared vessel known as 'The Great Galley'.

The ship's genesis

Henry Grace à Dieu was constructed at Portsmouth on conventional lines though on an unusually large scale, as a four-masted carrack, with a deep waist and reinforced deck in which the larger guns were set.

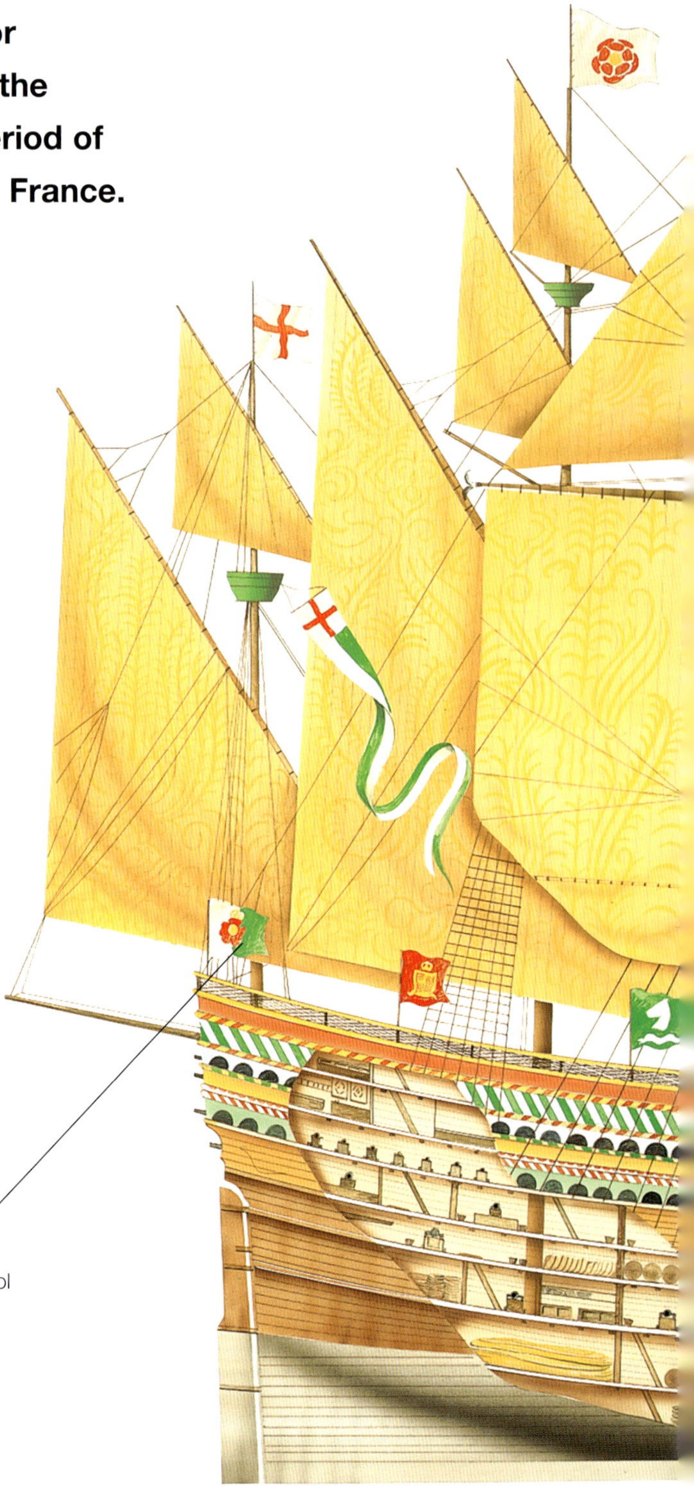

Heraldry
Heraldry played an important part in conveying the status of a warship and its commander. If the King were on board the royal standard would fly from the main mast. The red Tudor rose was a national symbol at the time.

Henry Grace à Dieu had seven decks, with larger guns mounted on two. As a flagship it had to have ample accommodation for senior officers and their entourage, and also for the King himself.

Rig
The topgallant sail was introduced around the turn of the fifteenth century in answer to the need for more power to drive deeper and heavier ships. Though three-part masts were now quite usual, the upper yards were narrower and lighter than the mainyards.

Sail
A painting of 1520 shows the ship's sails as yellow (probably a gold effect was aimed at), and while this was probably a specially prepared suit of sails for a special state occasion, taking Henry VIII to France, 'Great Harry' is usually portrayed in this way.

Gun decks
With two main gun decks and lighter guns mounted above, this was a very formidable ship for its day, even though it is doubtful if its gunners could manage the co-ordinated broadside of later times.

Early warship guns were of lightweight construction, barrels made in sections, and mounted with rope restraints on shaped wooden blocks. They were muzzle-loaded, with the rammer to push the ball down against the powder charge.

This positioning was necessary to provide stability, helped by the concentration of weight, but it also made it easier to train these guns at the waterline of enemy vessels, at a time when ships fought at close quarters. The development of the watertight gun-port was quite recent, and made a significant difference to the tactics of naval battle with the ability to fire a broadside at low level.

The other guns, of lighter calibre, were intended to shoot at the rigging, decks, and fighting tops of hostile craft, and so were mounted on the forecastle, four decks above the waist, and the poop deck (two decks above), in order to get the advantage of height. In all there were seven deck levels above the keel, providing room for stores and cannon balls, as well as space into which soldiers could be crammed for relatively short cross-Channel journeys. Topgallant sails, on topmast yards mounted above the upper fighting-tops, were relatively recent, and 'Great Harry' carried these as well as upper and lower lateen sails.

The long poop contained apartments suitable for the King and his senior courtiers and officers. Soldiers and sailors lived in the forecastle, in a men-forward, officers-aft pattern that would endure in the Royal Navy into the age of iron and steam.

Specification

Dimensions	Length 57.9m (190ft) x 15.2m (50ft) Displacement c1360 tonnes (c1500 tons) burthen
Rig	4 masts, square rig with 2 lateen masts
Armament	(1545): 21 bronze cannon, 130 iron guns, 100 hand guns
Complement	c700

Manpower

The crew was divided into sailors, responsible for operating the ship, soldiers, for fighting, and gunners.

While the breakdown of *Henry Grace à Dieu*'s crew is not known, the smaller *Mary Rose* had 200 sailors, 185 soldiers and 30 gunners. On the larger ship, with around 150 guns compared to 96, the number of specialist gunners would have been considerably greater.

There was no concept of a national war fleet at this time, and 'Great Harry' in its role as flagship was a symbol of the nation. In the event of a naval war, it would have led a hastily-assembled fleet of armed merchant ships, as happened in 1545 and again a generation later in 1588. But already we see the big warship as an instrument of national policy.

Limited action

It was not the biggest ship yet built in England: more than a hundred years before, the *Grace Dieu*, built at Southampton in 1418, had an estimated displacement weight of 2494 tonnes (2750 tons). *Henry Grace à Dieu*, launched in 1514, saw little in the way of action in its first decades. In 1536–39 it was given a virtual rebuild, along with *Mary Rose*, in which its height was reduced and its weight lightened, and 21 new heavy bronze cannon were mounted.

It was not involved in action until 18 July 1545, when an invasion fleet of 235 French ships, many of them galleys, attacked Portsmouth, already England's prime naval base. The vessel came under fire, but a northerly wind helped it to bear down on the smaller and lower enemy vessels. It was at this critical time that the consort vessel, *Mary Rose*, capsized and sank, due to water entering through inadequately secured gun-ports. The engagement was an indecisive one, but the French invasion attempt was a failure.

Henry VIII died in 1547 and the ship was renamed *Edward* for his successor, King Edward VI. Though kept seaworthy, it saw no further active service. On 23 August 1553 it was destroyed by fire, at Woolwich Naval Arsenal in London.

The carrack

Carrack was a name given in England to a type of ship known in Italy as a cocha, with a rounded hull shape and square sails, carvel-built, with the planks laid edgewise, not overlapping. The flat stern was a usual feature, as was the massive forecastle built out over the bow. Originally single-masted, by the late fifteenth century carracks were commonly three-masted, with a triangular lateen sail on the third mast. A fourth mast, or bonaventure, was fitted only on the largest ships, and the topmast shown on *Henry Grace à Dieu*'s bonaventure in a 1545 drawing may be a decorative embellishment rather than a reality. Hooks were fitted to the ends of the mainyards to assist in grappling, and a large grapnel was slung from the crown end of the bowsprit for the same purpose. A stout net was slung on each side of the well from the gunwales to the centre walkway, to prevent attackers jumping on board.

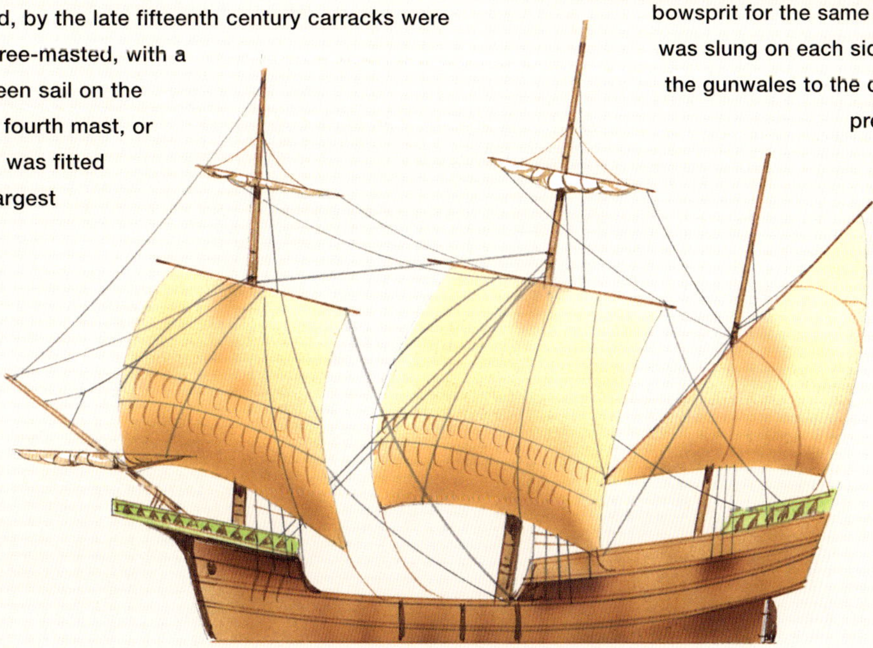

San Martín (c1579)

Though built as a Portuguese man of war, *San Martín* achieved fame as a 'Spanish galleon', the victorious flagship in the Battle of Terceiro (1582) against France, but less fortunate as flagship of the Great Armada sent against England in 1588.

In 1580 Portugal was annexed to Spain and its naval strength was combined with Spain's, including some fine ships of the galleon type. The Portuguese naval tradition was a strong one and the clean, simple lines of *San Martín*, with its two gun decks, exemplify the best warship design of 1570–80, and make an interesting comparison with those of 'Great Harry', which was still something of a floating castle rather than a fighting ship.

San Martín was a fine example of Portuguese shipbuilding skills. Note the bonaventure mast, set well in-board and not requiring the precarious stern-boom fitted on *Henry Grace à Dieu*.

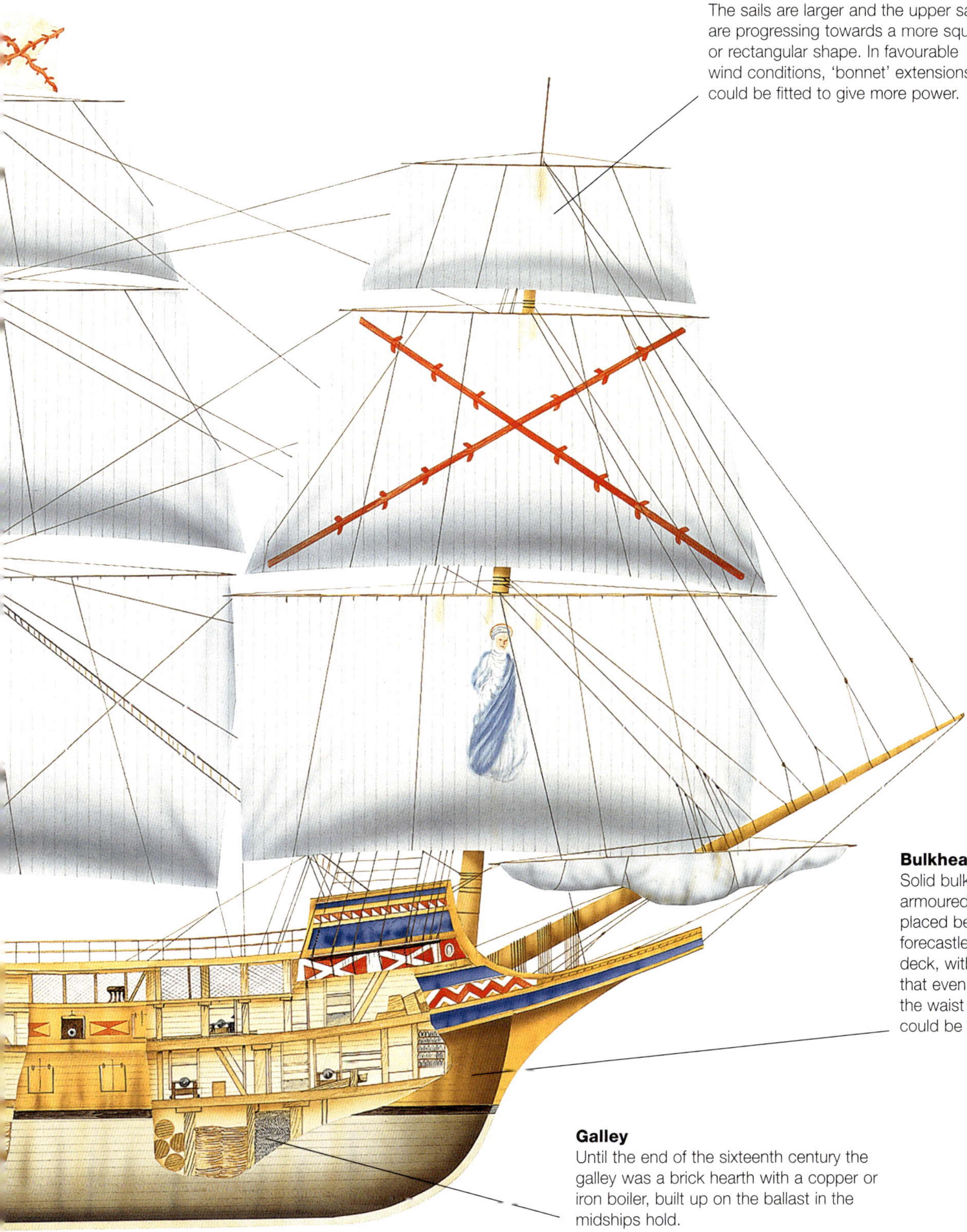

Rig
The sails are larger and the upper sails are progressing towards a more square or rectangular shape. In favourable wind conditions, 'bonnet' extensions could be fitted to give more power.

Bulkheads
Solid bulkheads and armoured doors were placed between the forecastle and the half deck, with firing loops, so that even if an enemy held the waist of the ship, they could be raked with fire.

Galley
Until the end of the sixteenth century the galley was a brick hearth with a copper or iron boiler, built up on the ballast in the midships hold.

Specification

Dimensions	Length 37.3m (122ft 3in); Beam 9.3m (30ft 9in) Displacement 907 tonnes (1000 tons)
Armament	48 heavy guns, plus light pieces
Rig	3 masts, square rig (possible 4th stern mast)
Complement:	350 seamen and gunners; 302 arquebusiers and musketeers

Evidently *San Martín* was better than anything Spain currently had, as it was very quickly given the status of *capitana general*, or flagship. On 15 July 1582 it led the fleet at the Battle of Terceiro, off the Azores, when the Spanish, commanded by the Marquis of Santa Cruz, defeated a 60-ship French fleet. This was the first fleet action between ships of the galleon type, and 10 French ships were sunk without claiming a single Spanish one.

Spanish Armada

Carrying so many guns, *San Martín* was purely a fighting ship, and had no further active deployment until May 1588, when a huge fleet was assembled for the invasion of England. Again *San Martín* was the headquarters ship, carrying the Duke of Medina Sidonia and his staff. The story of the Armada's failure, against a combination of English tenacity and ill weather, is well-known. In the context of the history of the battleship, *San Martín* provides a lesson which may not have been so apparent at the time. The Spanish galleons were warships intended to engage in close combat and their guns were not effective at other than close range. Although generally smaller, the English ships had guns of greater range and could bombard the Spanish from a distance without risking being boarded by enemy soldiers.

San Martín, in an hour-long duel with the *Ark Royal* on 1 August and another with *Triumph* on the 4th, was holed beneath the waterline and was rescued by the intervention of other ships. After a few days partial respite, battle was joined again on the French side of the English Channel from 8 August, off Gravelines, and *San Martín* fought a fierce rearguard action against numerous English ships, including Sir Francis Drake's *Revenge*, while the rest of the fleet moved northwards along the Dutch coast. This was a close action, but not so close that the galleons could grapple their fast-moving opponents, board them and force them to surrender. By 9 August it was obvious that the Armada was not able to command the sea and bring its invasion force to land. Its return to Spain round the stormy coasts of Scotland and Ireland left many ships wrecked. *San Martín* was one of the 67 which got home, out of over 130 ships, reaching Santander on 23 September.

The lesson was that the quality of guns and gunnery was important: not merely an extra but potentially decisive in a battle on the open sea. This was not necessarily palatable to seamen brought up in the old grappling and boarding tradition. The technology of cannon casting, and of the missiles fired – solid iron balls – was to change only slowly and gradually, and two hundred years later ships armed with far more and heavier guns would still seek to get alongside an opponent and storm its decks with armed men. But the value of cannon had been clearly shown.

Size, strength and manoeuvrability

Ark Royal, the flagship of the English Admiral, Lord Howard, was of 629 tonnes (694 tons) burthen compared to *San Martín*'s 907 tonnes (1000 tons), but carried four masts and 38 guns, including four 60-pounders, four 30-pounders and twelve 18-pounders. Like most of the other English ships, it was more manoeuvrable than the bigger Spanish vessels,

The galleon

The galleon was a large ship, typified by a narrow length-to-beam ratio, a lower freeboard (deck height above waterline) than was usual and a square stern. It was not built as high at forecastle and sterncastle as preceding ships, and had three, occasionally four, masts. The deck extended in a pointed beak over the bow. The aim was to produce a fighting platform that was faster and easier to work than the high-built warships typical before 1575. Gun-decks were built-in and usually held the latest models of cannon. The English *Revenge* (1577) carried 2 demi-cannon, 4 cannon-periers, 10 culverins, 6 demi-culverins and 10 sakers, apart from smaller guns. All the major seafaring countries built galleons: England, France, Holland, Spain, all with similar features but varying in size, type of armament and even rig: the term (first used in English around 1529) is not a precise one in ship-design.

This painting by Charles Dixon (1872–1934) of the action between *San Martín* and *Ark Royal* in the Channel is indicative of how the English ships were smaller and handier than the massive Spanish vessels. Both were the flagships of the combating fleets.

but despite superior gunnery, the English fleet did not have the power to destroy the massive galleons. Size and strength counted for something, too. After September 1588, *San Martín* must have been in poor condition, badly in need of repair and refitting. The war was not over, and the Spanish had to regroup and reform their naval forces against attack from England, France and the Netherlands.

All effective craft must have been kept in action, or quickly restored, but *San Martín*'s subsequent fate is unknown

Vasa (1628)

Vasa had the shortest career of almost any warship. Built by the Royal Dockyard in Stockholm in 1628, the 64-gun vessel sank on its very first voyage, only a few minutes after departure.

The discovery of *Vasa*'s wreck, its recovery and preservation, have made it the longest-lived of seventeenth-century men of war, a treasury of artefacts and information relating to naval technology and life at the time. Vasa, or Wasa, was the surname of the Swedish kings, and the ship was commissioned for one of the greatest, Gustavus Adolphus, who in 1628 was

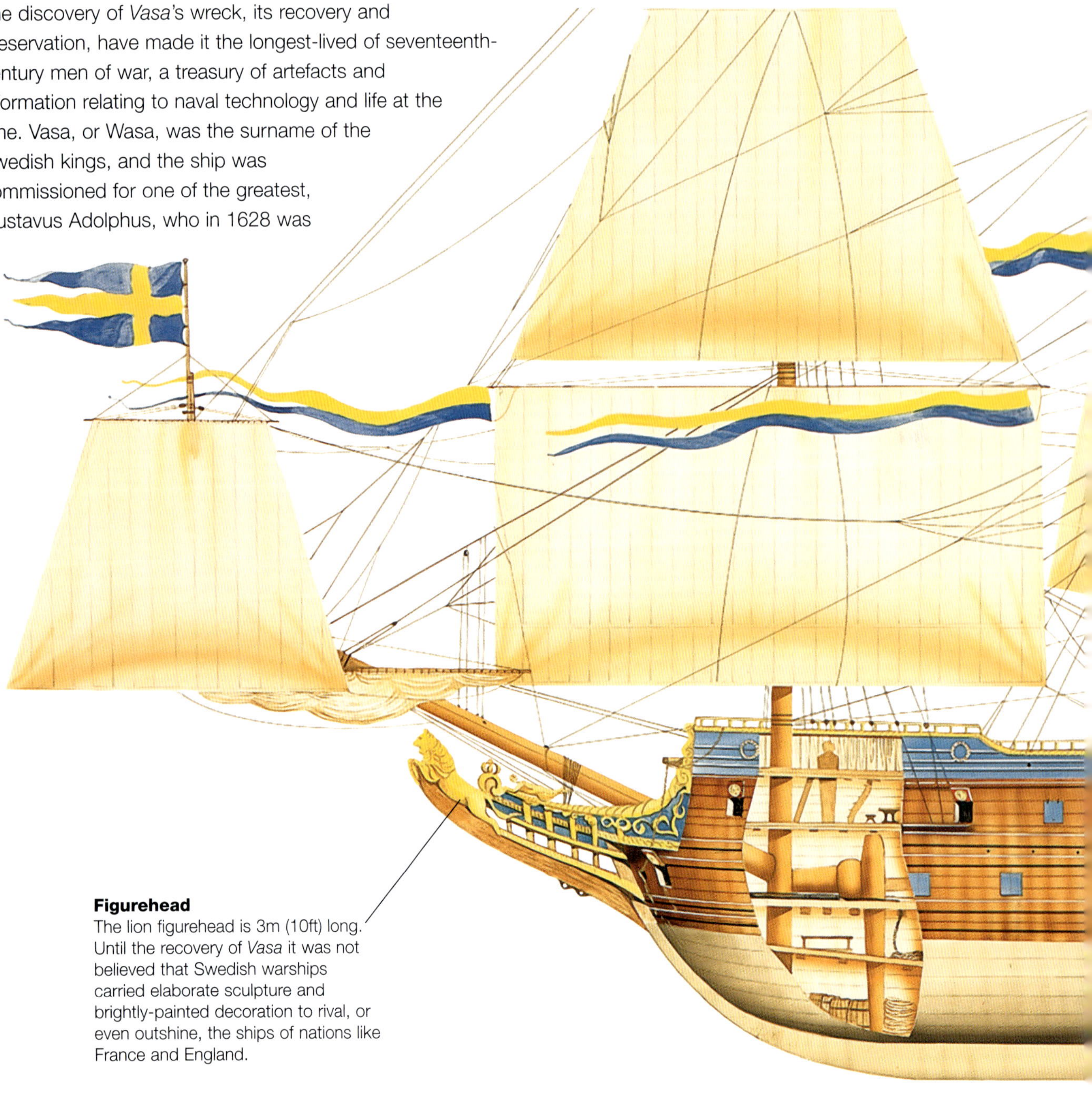

Figurehead
The lion figurehead is 3m (10ft) long. Until the recovery of *Vasa* it was not believed that Swedish warships carried elaborate sculpture and brightly-painted decoration to rival, or even outshine, the ships of nations like France and England.

Spritsail topsail
In the early seventeenth century a mast was built up from the tip of the bowsprit to carry an extra sail in addition to the spritsail carried below the bowsprit.

Gun ports
These had been in use for over a century and the danger of leaving the lowest level open was well-known. *Vasa's* were of very solid construction, set in strong frames, upwards-hinged, and with a narrower inner layer to seal the sides effectively.

Guns
Vasa could fire 267kg (588lb) of shot from the guns of one side but tactics still centred on the firing of individual guns in action against a single ship, rather than the line-of-battle broadside actions still to come.

The hull of the *Vasa* is preserved in a specially-constructed museum building, with a temperature- and moisture-controlled environment to keep the timbers from drying out. Archaeological work on the ship and its contents is still ongoing.

deeply involved, with his army, in warfare in Germany.

As one aspect of the wide-ranging Thirty Years' War (1618–48) Sweden was at war with Poland, and mounting a naval blockade of the city of Danzig (Gdansk). *Vasa*'s first mission was to join this blockading force. On the first stage of its journey from Stockholm on 10 April 1628, there was a party atmosphere, with women and children as guests on board for the trip out of the bay. The wind was so light that *Vasa* was warped out of the harbour, winding itself along an anchored line dropped from a longboat. Sail was set as occasional gusts of wind came, and one of these suddenly laid the ship over, allowing water to flow in through open gun-ports, and it promptly sank, drowning some 50 of the 250 people on board.

Raising the ship

Salvage efforts followed, and most of the guns were retrieved in 1663–64. But it was only in the 1950s that it became technically possible to raise the hull, lodged in mud at a depth of 35m (115ft), and on 24 April 1960 it was brought to the surface and transferred to a dry-dock and then to a specially-constructed building. The ship had been designed by a Dutch builder, Henrik Hybertson de Groot (the Dutch were celebrated ship-builders). Just like *Henry Grace à Dieu*, a hundred years earlier, it was not just a fighting platform, but a proclamation of national and royal prestige and splendour. Sculptures and carvings adorned it in profusion, on classical and Biblical themes intended to reflect glory on the house of Vasa, and gilded or brightly painted. A great range of artefacts necessary to stock and operate a big warship were also recovered, ranging from entire sails, items of rigging, ropes and steering gear, to the clothes and personal property of the crew, wooden and pewter plates, drinking vessels, sewing kits, clothes packed in chests, belts and shoes. Over four thousand copper coins were found on board. The chill Baltic waters preserved most items except those of ferrous metal, which were almost completely rusted away.

Examining the wreck

Vasa was a three-masted ship, and examination of the wreck showed that the mainmast was raked back some 9 degrees and the deck had a considerable slope: proving aspects of contemporary design previously seen only in drawings and paintings. Despite the height of the hull, with a quarter-deck 20m (65ft) above the waterline, it carried only 109 tonnes (120 tons) of stone ballast, and the inherent instability of the vessel was suspected as soon as it was launched. But Sweden was short of large ships, having lost numerous vessels during the period of *Vasa*'s construction in storms and in the Battle of Oliwa (1627) fought in the Southern Baltic Sea against a Polish fleet; and with letters from the King insisting on early completion, the building and fitting out were done in a hurry. Hybertson had died in the early stages of construction, and there was mutual blaming between the naval authorities and the builders. The comparative narrowness of the hull in relation to its length was commented on, but the length had been set by the King, whom no one dared to criticise. To sail the ship with its lower gun-ports open, when instability had already been identified, seems a foolish action of the captain, Söfring Hansson, but he was not censured. In the end, a court of inquiry assigned no blame to anyone, and the disaster was treated as an 'Act of God'.

Specification

Dimensions	Length 69m (226ft), Beam 11.7m (38ft 3in), Draught 4.9m (15ft 4in)
Displacement	6682 long tons (6789t)
Propulsion	3 masts, square rig on fore and main, lateen mizzen
Armament	48 24-pounder, 8 3-pounder, 2 1-pounder, 1 16-pounder, 2 62-pounder, 3 35-pounder guns
Complement:	145 crew, 300 soldiers

Vasa's guns

Specifications for *Vasa*'s armament were changed during the years of construction, and a variety of English names, translated from Swedish, have been applied to the different types of gun. *Vasa* was probably the best-armed warship of the period, although modern analysis suggests that their bronze was not of the best quality. In 1626, 72 24-pounder bronze cannon were ordered for it, and it is likely that this number required revision of the design to build a second gun deck not allowed for in the original plan. Eventually *Vasa* carried only 64 carriage-mounted guns, with 48 24-pounders, arranged in 24 on each gun deck, 10 lighter cannon and 6 guns of mortar type, firing large explosive shells ranging in weight from 7.25 to 28kg (16 to 62lb). At a time when men of war tended to carry a variety of guns, some of them having seen service in successive ships, *Vasa*'s artillery was of relatively standardised type. During 1663–64, 53 guns, weighing up to 1.36 tonnes (1.5 tons) each, were salvaged from the wreck, and sold in Germany.

Vasa's legacy

Vasa had been intended as the first of five capital ships. Four others of similar design, but of longer life followed: *Applet*, *Kronan*, *Scepter* and *Göta Ark*. These served satisfactorily into the 1660s, and it may be assumed that some modifications were made to improve their stability (*Applet* had a slightly broader beam of 13.18m (43ft 3in). Though Sweden was a comparatively rich country, the cost of 40,000 thalers for each ship was a heavy one.

Sovereign of the Seas (1637)

The most ornate ship ever built for England's Royal Navy, the first English warship to have three full gun decks, the first to carry 100 guns and the most expensive yet built – the *Sovereign* claimed a range of superlatives.

In every way a 'King's ship', the vessel was ordered in 1634 by King Charles I as part of a very proper concern to improve and enlarge the Navy at a time when other nations, notably the Dutch and the French, were developing their own overseas trade and building new ships. The designer was Phineas Pett, England's leading shipbuilder, and it was constructed at the Woolwich Dockyard, London, by his son Peter.

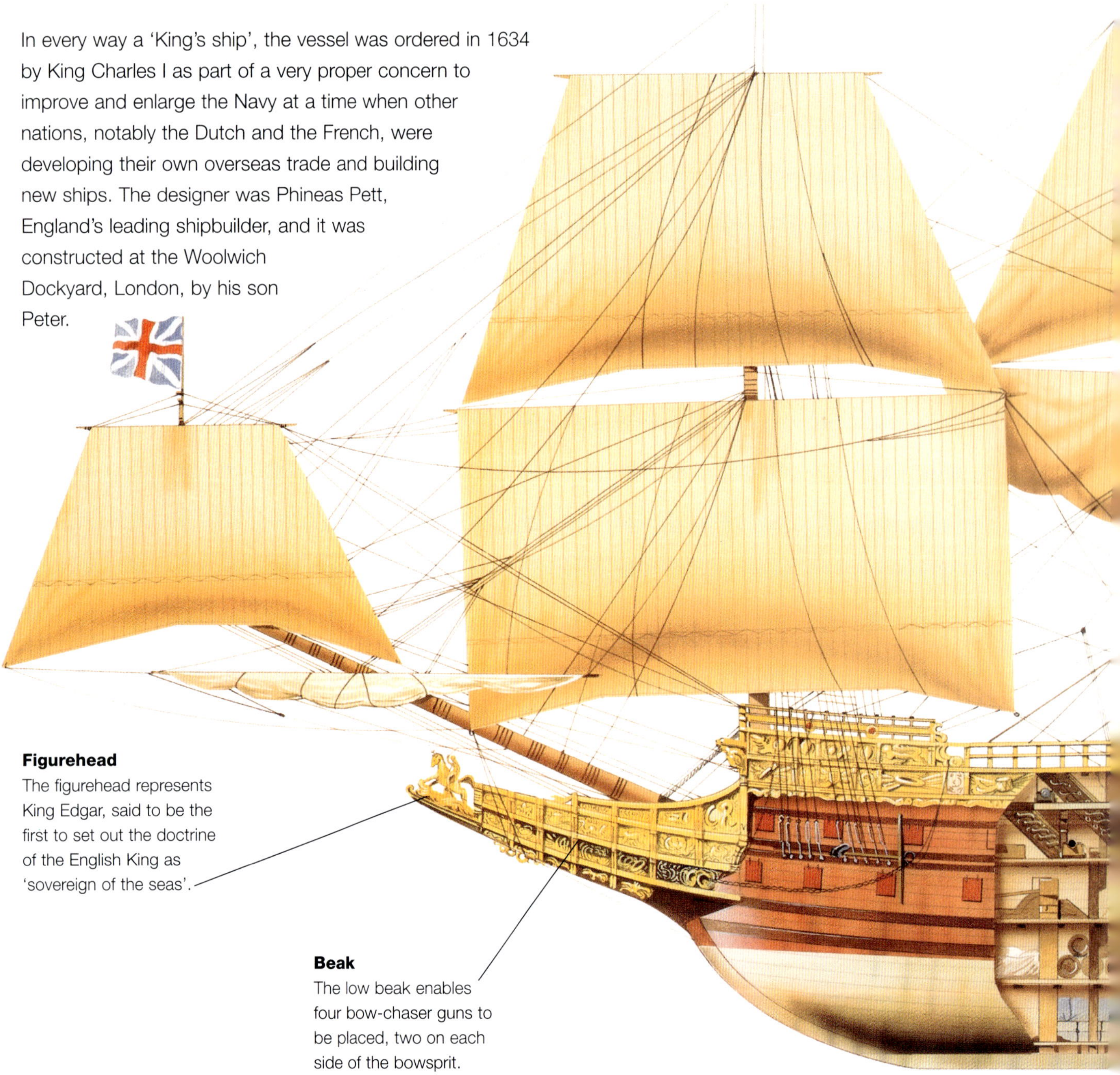

Figurehead
The figurehead represents King Edgar, said to be the first to set out the doctrine of the English King as 'sovereign of the seas'.

Beak
The low beak enables four bow-chaser guns to be placed, two on each side of the bowsprit.

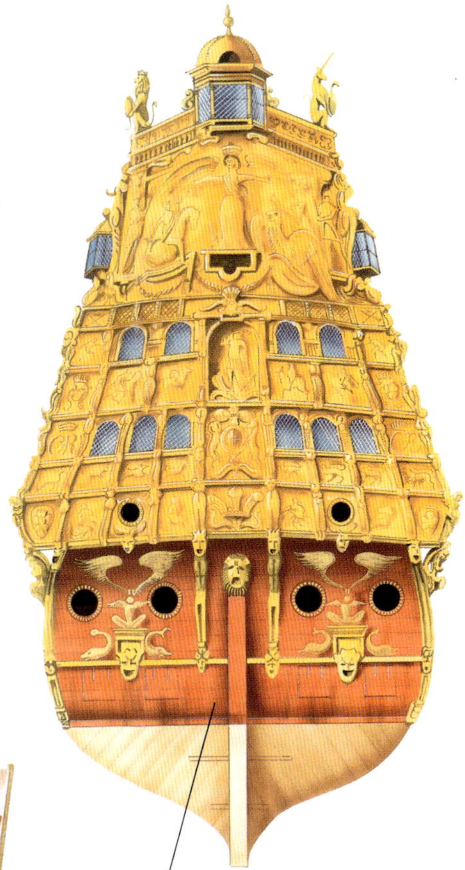

Rig

The vessel mounted 13 sails including royals above the topgallants on fore and main masts, and mizzen topsail and topgallant. The rigging includes martnets to help secure the upper edges of the foresail and main topsail.

Stern

The stern is rounded at the waterline, flattening out around 3m (10ft) above it in a manner typical of later British men of war.

Galleries

The walkways have been elaborated into decorative windowed galleries, as extensions of the main cabins.

Three masts, rather than four, became the standard for the English man of war. Additional sails, a larger sail area, and stout canvas with hemp rigging, together with good underwater lines, ensured speed.

Key developments

With the increasing emphasis on gunnery, certain changes in warship design and organisation appeared. The anti-boarding net across the well was no longer used. Parts of the quarter-deck and forecastle deck were floored with gratings rather than planking, to allow powder fumes to disperse from below. The whipstaff, an upright lever attached to the tiller, made steering somewhat easier and enabled the helmsman to stand in a raised position from which he could watch the sails. Larger capstans, centrally mounted, helped in drawing up heavy anchors. The wheel, for steering, did not come into use until the beginning of the eighteenth century. From the early sixteenth century shipbuilding was increasingly scientific and textbooks began to appear. As a result, innovations were passed on quickly and hull design and rig were increasingly similar in the different navies of Europe. But British men of war continued to ride lower in the water than their French and Spanish counterparts.

Phineas Pett had already built a man of war that was partially three-decked, the 55-gun *Prince Royal* of 1610, but this ship was half as big again. Whether the triple gun decks were the King's idea or Pett's, both liked the thought, the one as a challenge to his skill, the other as a symbol of his royal power.

The King levied a special tax, 'ship money', on his reluctant subjects to pay for his naval policy. *Sovereign of the Seas* was to cost £65,586, a colossal sum for that period. The ship's name was deliberately chosen: an old tradition, which Charles I's government tried to revive, held that the King of England held command of the surrounding seas – a claim which neighbouring nations rejected.

Armament

A contemporary description noted that the ship had three flush decks, a forecastle, a quarter-deck and a roundhouse. On the lowest gun deck were 30 ports, for 30-pounder demi-cannon and whole cannon; on the middle deck were 30 ports again, for demi-culverin 10-pounder and whole culverin. The upper deck had 26 ports for smaller guns, and there were 12 ports on the forecastle, and 14 on the half-deck. The report also mentions 13 or 14 ports for 'murdering pieces' – these were scatter-shot guns that could sweep the deck of an adjacent enemy, and many loop holes for musket shots. These were a recent addition to warship design, since the development of the musket. In addition, the ship carried 10 pieces of 'chase ordnance': guns mounted in the bows and at the stern intended to extend the ship's range of fire to 360 degrees.

This was a 'state of the art' ship which deliberately pushed out the frontiers of naval design and put new demands not only on its crew but on the dockyard support necessary to maintain such a large and heavily-gunned ship. Its bulk and weight required a considerable spread of sail to drive it forward at any speed: in this it was typical of other ships of similar size, but these were almost all merchant vessels, while a warship had to be faster than its prey. Sail design and technology was being improved, and *Sovereign* was one of the first ships to carry a fourth set of sails, royals, mounted above the topgallants, but even so it was considered a slow and unwieldy craft and it had difficulty in keeping up with a fleet formed mostly of smaller ships. In 1642 the number of guns was cut back to 90.

Specification (as built)

Dimensions	Length 39m (127ft) (Keel)
	Beam 14.17m (46ft 6in)
	Burthen 1380 tonnes (1522 tons)
Rig	3 masts, square rig with royals (after 1685)
	Complement: not recorded
Armament	102 cannon

Eleven anchors were carried, though only one was a main anchor, reported as weighing about 2 tonnes (2.2 tons).

Distinguished career

Sovereign of the Seas had a long active career. Following the establishment of the British Commonwealth it was renamed *Sovereign*, and much of the decorative carving was removed, taking away a good deal of dead weight. For a time it was the flagship of one of England's greatest admirals, Robert Blake. After the restoration of kingship in 1662 the 25-year-old ship was extensively rebuilt and *Royal* was prefixed to *Sovereign*. Its active service was mostly in the North Sea and the English Channel, in the course of three Anglo-Dutch wars, fighting in the Battles of Kentish Knock (1652), Orfordness (1666), Solebay (1672) Schoonveld (1673) and the Texel (1673). Another large-scale rebuild was done at Chatham in 1685, and *Royal Sovereign*, by now a venerable vessel, survived to fight at Beachy Head (1690) and Barfleur (1692).

Throughout this time it was the largest ship in the English fleet and its firepower (restored to 100 guns in 1660) made it a powerful asset in battle. It was laid up at Chatham naval dockyard, where on 27 January 1697 a fire broke out on board and burned it to the waterline.

Sovereign of the Seas **under full sail, a painting based upon John Payne's contemporary engraving of 1637, 'To the great glory of our English Nation, and not paraleld in the whole Christian World.'**

Le Soleil Royal (1669)

Another floating monument to royal pretension, but also an extremely powerful fighting ship, carrying over 100 guns, the ship typified the regeneration of France's Navy carried out by the minister Jean-Baptiste Colbert.

Classified as a *vaisseau* (first rate), as was typical for a first-rate man of war (the term battleship did not come into use until around 1884, as an abbreviation of 'line-of-battle ship' which was first recorded in 1705), *Le Soleil Royal* was splendidly adorned with colourful carving, the hull painted black, white and blue with gold embellishments, and at the stern was a massive sculpture in wood of the chariot of the sun drawn by winged horses. The name was a reference to the 'Sun-King', Louis XIV, replacing the original plan to call it *Grand-Henry*, in memory of one of his predecessors, Henri IV.

France needed new warships at this time, and in fact several perfectly serviceable smaller ships could have been constructed for the price paid for *Le Soleil Royal*, but that was not how the minds of seventeenth-century strategists worked.

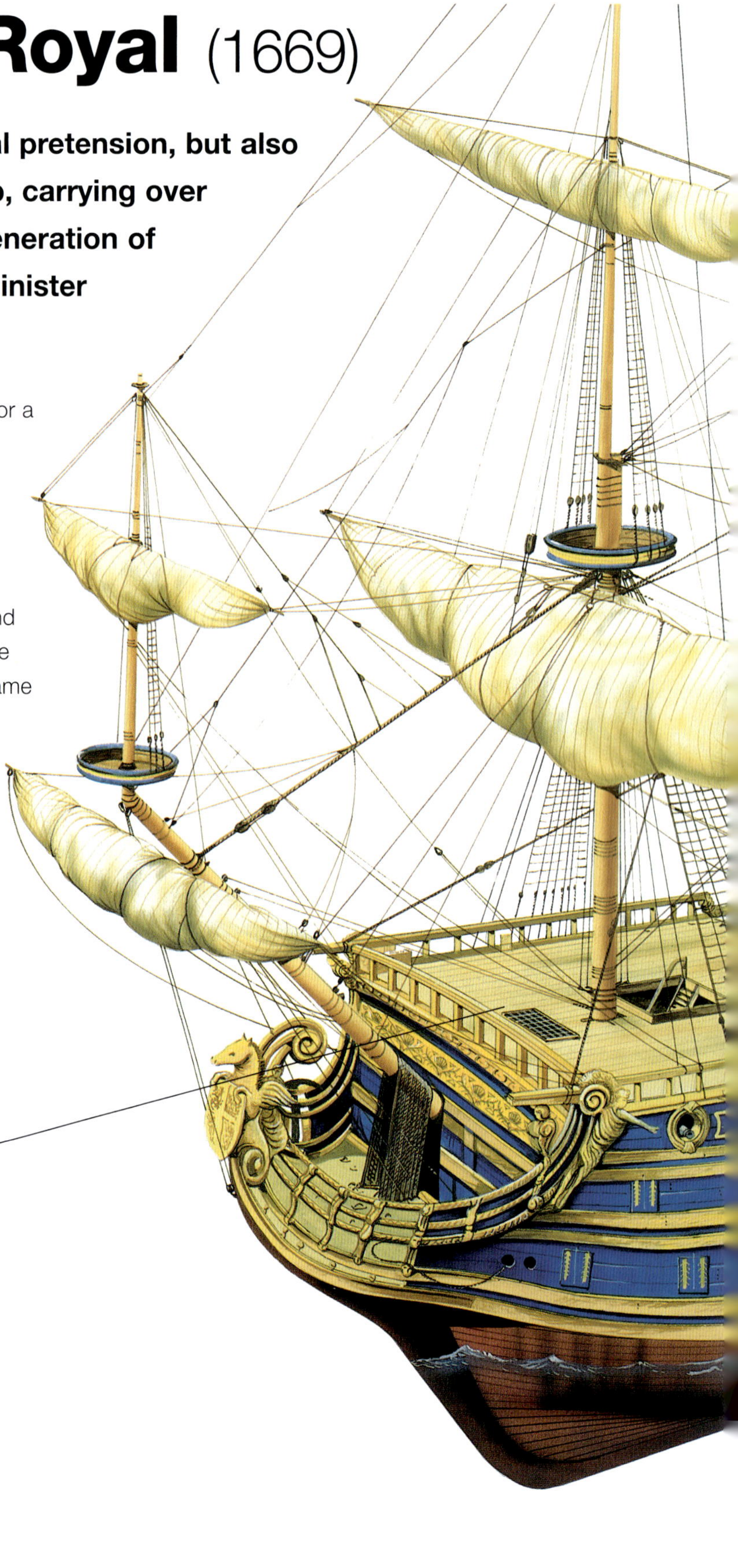

Decks
The half-deck and quarter-deck rise to the poop deck. Etiquette, discipline and status determined who might be allowed to set foot on each deck, and when.

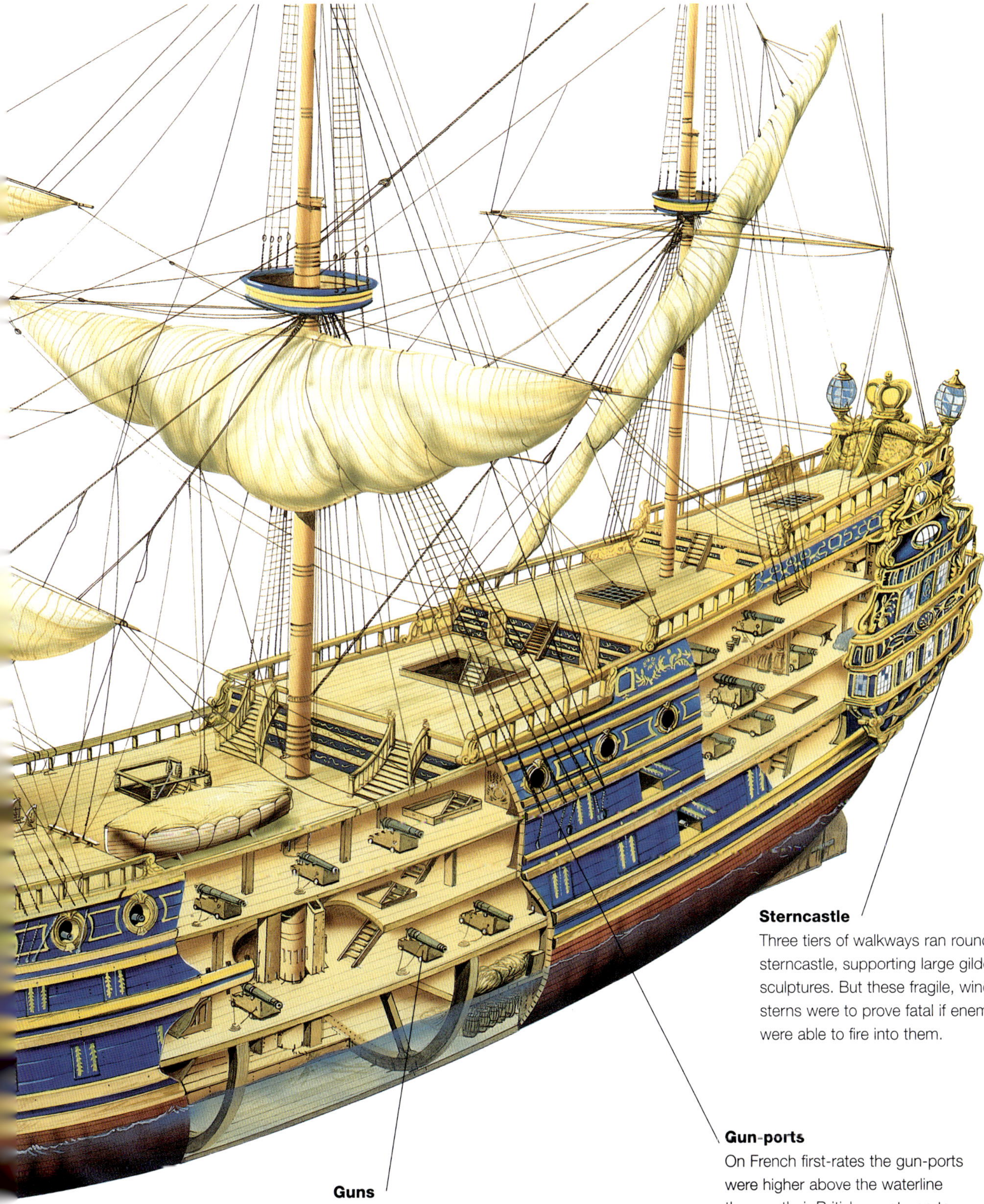

Sterncastle
Three tiers of walkways ran round the sterncastle, supporting large gilded sculptures. But these fragile, windowed sterns were to prove fatal if enemy ships were able to fire into them.

Gun-ports
On French first-rates the gun-ports were higher above the waterline than on their British counterparts. Wider in the beam, the ships were more stable which improved thoir fighting ability.

Guns
The quincunx or staggered formation of the guns can be seen, a positioning that helped balance and also relieved undue stress on the ship's jointwork.

Le Soleil Royal

A modern watercolour of *Le Soleil Royal* emphasises the decoration lavished on a French first-rate in the age of the Sun-King, Louis XIV, as well as indicating the ship's great length.

The vast, gilded, needlessly expensive showpiece was considered essential for national prestige. Perhaps also it contributed to naval morale. The ship was built at Brest by the master-builder Laurent Hubac and launched on 13 December 1669. With guns installed, it was commissioned in August 1670. France was Europe's most formidable land power, but despite its three coastlines, the Navy had been neglected. Two divisions were required, one for the Mediterranean Sea, formed primarily from galleys and other oared ships, and one for the northern coast, fronting England and next to the increasingly powerful and

Specification

Dimensions	Length 55m (180ft 5in), Beam 15.6m (51ft 2in), Draught 7.6m (24ft 11in), Displacement c1088 tonnes (c1200 tons)
Rig	3 masts, full ship rig
Armament	98 to 104 bronze cannon, at different times
Complement	1200

aggressive Dutch states. The French had used galleys in Channel warfare before, but it was now appreciated that larger ships, better able to withstand storms, and more heavily armed, were essential to counter the English and Dutch fleets.

Heavy cannon

By now it was quite accepted for a first-rate warship to mount heavy cannon on three decks, as well as lighter guns on the forecastle and poop deck. Flagship status was indicated by the large triple lanterns at the poop and another fixed to the mizzen-mast. Its cannon were of bronze rather than cast iron, mounted behind gun-ports painted bright red. In its early years it appears to have been little used, but in 1686 it went through an extensive refit, and in 1688 the number of guns was increased to 104.

In 1688 it became the flagship of the Count of Tourville, and when Louis XIV declared war against Spain and England in 1689, Vice-Admiral Tourville led a fleet of 75 ships which defeated a combined fleet of 39 English and 21 Dutch ships, off Beachy Head on 10 July 1690 in what was known in France as the Battle of Cape Béveziers. By this time the concept of the 'line of battle' had replaced the earlier, less ordered form of a naval battle, and the French fleet was formed into three divisions, with *Le Soleil Royal* leading the central division. From then on it was in regular action.

Tourville saw his fleet reduced when 15 ships were transferred to the Mediterranean. Though summoned back to Brest, they were delayed and did not form part of the fleet which was ordered to clear the Channel of enemy craft in preparation for an invasion intended to restore King James II to the English throne. With *Le Soleil Royal* again as flagship, the French with 44 ships and 11 fireships fought an Anglo-Dutch fleet of double its number off Barfleur on 20 May 1692. Tourville broke off the action and tried to bring his ships home, but the fleet, pursued by the victors, became divided. *Le Soleil Royal* suffered heavy damage and more than half of the ship's complement of 973 men were killed or wounded; and on 30 May Tourville shifted his flag to *L'Ambitieux*.

Set ablaze

Under its commander, Desnos, the stricken *Soleil Royal* was forced eastwards by the tide, to run aground off Cherbourg, with two other ships. Still under attack, and returning fire, the crew cut down the masts in the hope of getting the ship further inshore, while the English prepared fireships to drift in. On 2 June, a fireship struck against the French giant and soon *Le Soleil* was ablaze. The powder magazine exploded, killing all but one of the remaining crew. Its destruction, and that of 13 others, was hailed in England and Holland. Medals were struck. But the wealth and resilience of France meant that within a year, the Channel was again being contested. A second *Soleil Royal*, less magnificent, was launched in 1693.

First-rate

A royal ordinance defined the characteristics of a French first-rate, to ensure that ships would be built to the same specifications. The prime requirement was that it should have three full-length gun-decks. Only two ships, *Le Soleil Royal and Royal-Louis,* were to carry a 'castle' at the bow. The Royal French Navy had five rates of ship and in 1692 it had 22 first-rates (80–100 guns), 27 second-rates (60–80 guns), 46 third-rates (50–60 guns), 16 fourth-rates (40–50 guns) and 21 fifth-rates (20–36 guns). All these were deemed worthy of joining the line of battle. England's Navy, first to employ the rating system (from 1653), had six rates. Instead of a free-for-all, the fleets sailed in line, following flag signals, exchanging broadsides, each trying to get the other fleet placed at a disadvantage, so that they might be able to close in and complete the action. A successful admiral now had to be a skilful and sometimes bold tactician.

Victory (1765)

HMS *Victory* is the oldest warship still officially in commission, as flagship of the commander-in-chief at Portsmouth, though the role is ceremonial. Nowadays it is a museum ship and has been permanently in dry dock since 1922.

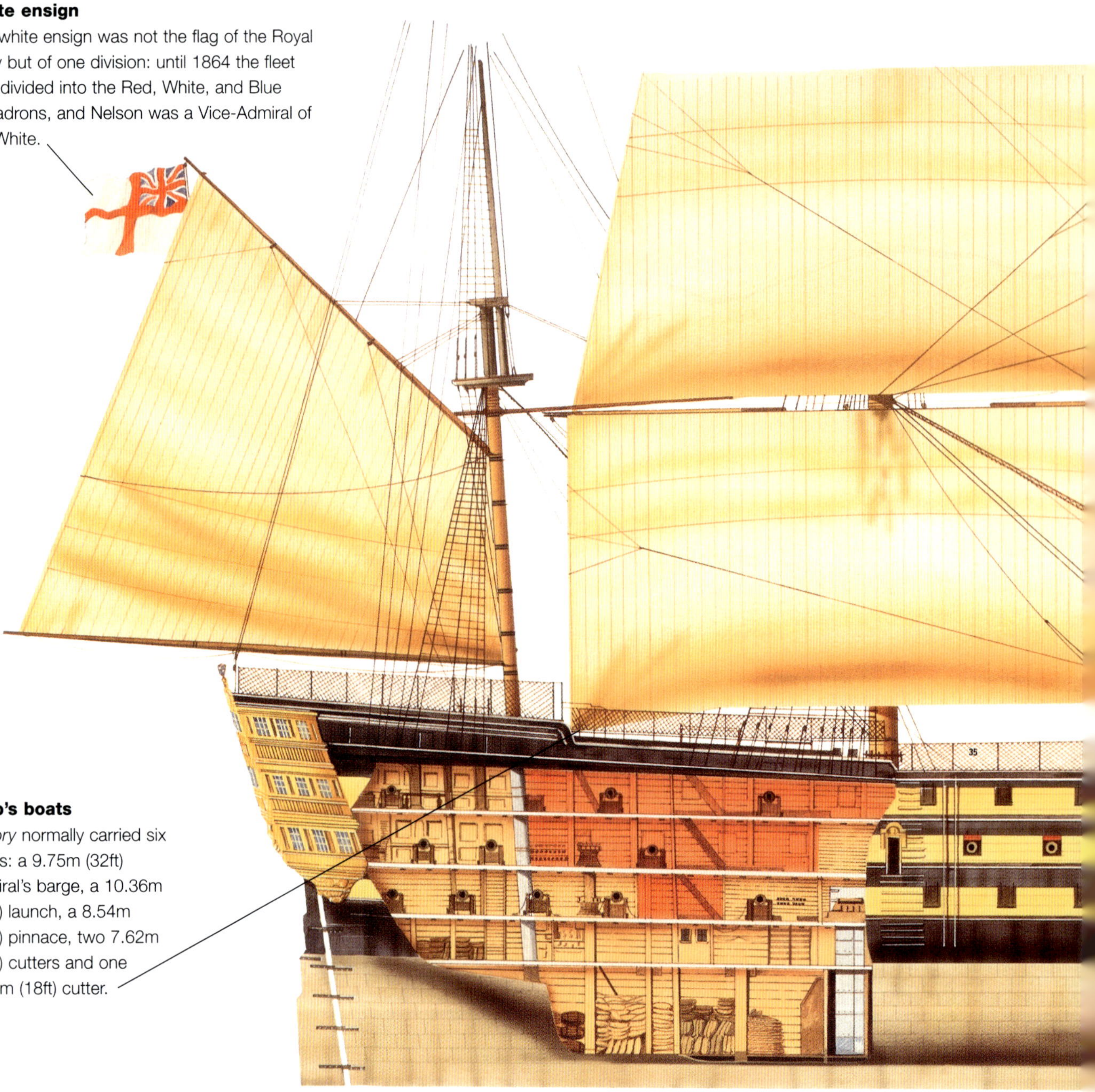

White ensign

The white ensign was not the flag of the Royal Navy but of one division: until 1864 the fleet was divided into the Red, White, and Blue Squadrons, and Nelson was a Vice-Admiral of the White.

Ship's boats

Victory normally carried six boats: a 9.75m (32ft) admiral's barge, a 10.36m (34ft) launch, a 8.54m (28ft) pinnace, two 7.62m (25ft) cutters and one 5.48m (18ft) cutter.

Hammocks
Round the open deck hammock netting is bundled against the railings, u-shaped iron fittings where the sailors' hammocks were stowed during the day. In action it formed a defence against light fire.

Rig
By *Victory's* time the spritsail topsail, with its vulnerable joint, was gone, but the vastly long bowsprit, 33.5m (110ft), carried yards for two spritsails as well as supporting the stays of two headsails. Altogether the ship carried 42km (26 miles) of rigging.

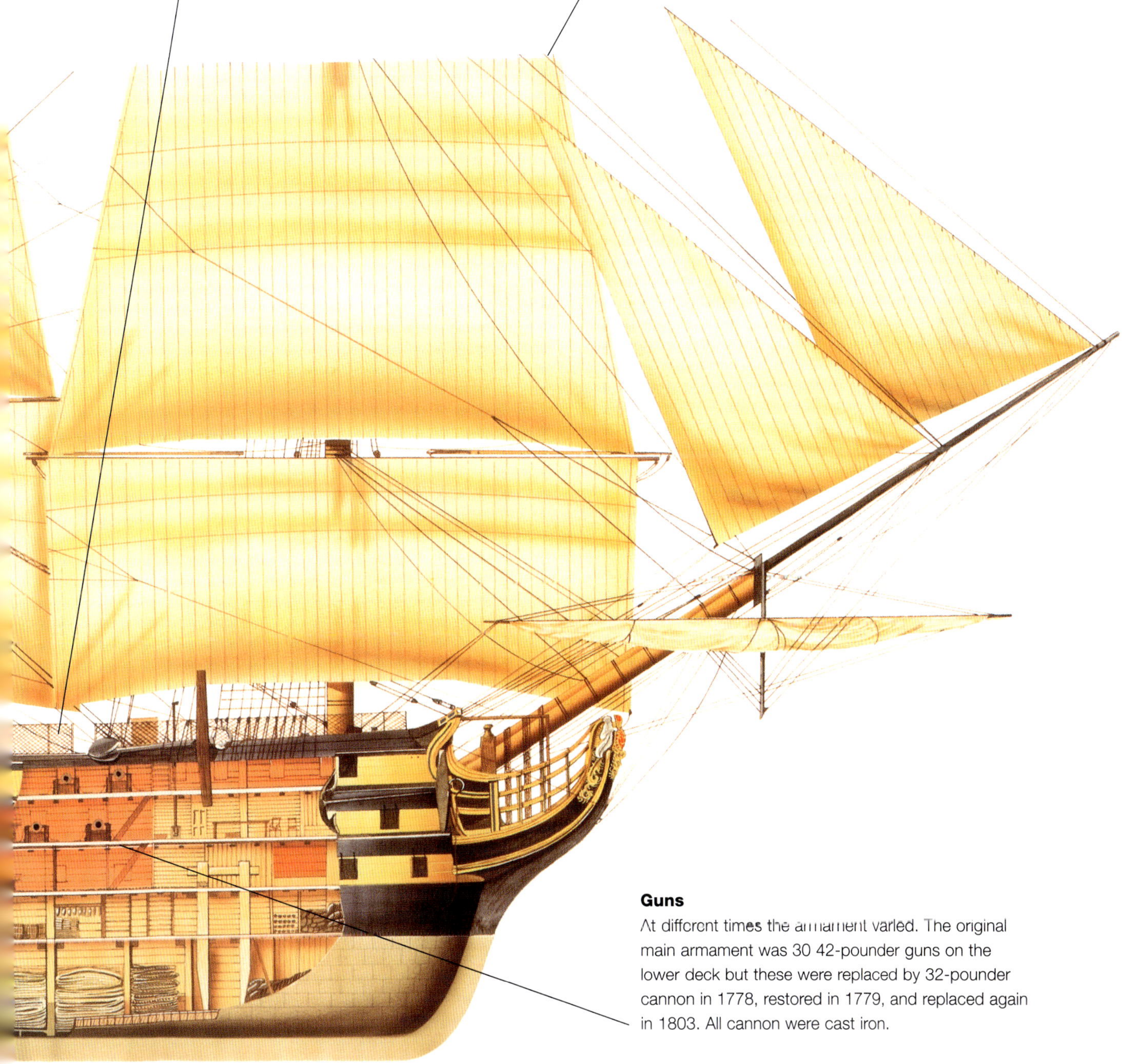

Guns
At different times the armament varied. The original main armament was 30 42-pounder guns on the lower deck but these were replaced by 32-pounder cannon in 1778, restored in 1779, and replaced again in 1803. All cannon were cast iron.

Laid down in 1759, launched in 1765, *Victory* was not commissioned until 1778, when war with France was renewed and the Royal Navy was hastily brought to full operational strength.

A first-rate, carrying (at the maximum) 102 guns, it was built at Chatham Naval Dockyard, as a flagship, with accommodation for an admiral and his staff. Its 'great cabin' was divided into three sections for the admiral's day-room, dining room and sleeping quarters. The perception of the capital ship had changed since the time of *Sovereign of the Seas*: it was far less of a glory symbol and far more of a great practical fighting machine. *Victory* carried little in the way of ornamentation though there still was decorative carving round its beak and poop. Another reason was that there were far more ships of the line, and economies had to be found in order to pay for them.

Slade's design

Victory, which cost £63,176, was built to the design of Sir Thomas Slade, Surveyor of the Navy, and like other ships of the line was of oak construction, with a thickness of 60cm (2ft) at the waterline, and a keel of elm. Copper plating, introduced in the Royal Navy in 1761, was applied in 1780 to protect the hull from shipworm. The ship was considerably wider at the waterline than at main deck level, with an inwards-curving 'tumblehome' form in profile, the heavy guns being in staggered positions from the centre-line, and alternate placings on each deck, rather than stacked right above one another. The foremast is vertical but the main and mizzen masts have a distinct rake. Full sail extent was 5440m^2 (6510yd^2), making ships of this type both faster and more manoeuvrable than older vessels. Originally the stern galleries were open.

Flagship

In July 1778 *Victory* went into action as flagship of the Channel Fleet, under Admiral Keppel, at the indecisive first Battle of Ushant, with 35 men killed or wounded. A further battle off Ushant, on 12 December 1780, resulted in the capture of 15 French merchant ships. In 1782 *Victory* was Lord Howe's flagship at the relief of Gibraltar, but with the end of hostilities was paid off and held 'in ordinary' at Portsmouth until 1791. From 1792–95 it was Vice-Admiral Hood's flagship in the Mediterranean, then of Admiral Sir John Jervis. On 14 February 1797, with 15 sail of the line, Jervis intercepted a Spanish convoy escorted by 27 major ships. The resulting Battle of Cape St Vincent was a triumph for the tactics of Jervis, passing between the two divisions of the Spanish fleet – and of one of his captains, the independently-minded Commodore Nelson on the 74-gun HMS *Captain*, who prevented the Spanish line from reforming. Intensive service had left *Victory* in poor condition and in 1798–99 it was relegated to the role of a hospital ship at Chatham, but between 1800 and 1803 a major refit was done, with new masts, emplacements for 104 guns, and much internal rebuilding.

The cost, at £70,933, exceeded the original construction cost but a new ship would have cost even more. On 11 April 1803 the renewed *Victory* was ready and on 16 May Nelson, now a Vice Admiral, appointed commander of the Mediterranean Fleet, hoisted his flag on board. The ship was repainted in the black and yellow 'Nelson check' it still retains.

Trafalgar

At first employed in blockading the French base at Toulon, later in pursuing it across the Atlantic and back again, *Victory* and the British fleet spent two active but frustrating years before the long-awaited encounter with the combined French and Spanish fleets took place off Cape Trafalgar on 21 October 1805. Nelson's battle plan, to cut the enemy's line of battle in two places, meant that *Victory,* leading the weather division, had to withstand 45 minutes of broadsides before bringing its own guns to bear. Passing immediately astern of the French flagship *Bucentaure, Victory* fired a triple-shotted broadside with devastating

Specification (1805)

Dimensions	Length 69m (226ft 6in), Beam 15.8m (52ft), Draught 8.76m (28ft 9in), Displacement 1961 tonnes (2162 tons) burthen
Rig	3 masts, full ship rig
Armament	30 32-pounder, 28 24-pounder, 44 12-pounder cannon; 2 68-pounder carronades
Complement	c850

effect, and ran against *Redoutable.* For a short time it seemed that *Redoutable* might board *Victory,* but HMS *Temeraire* came up on the other side and the French ship was caught between the two. By 14:30, *Victory,* severely battered, was out of the battle, which raged on; 57 crew were killed and 102 wounded, the most notable death being that of Nelson himself, hit by a sharpshooter from the rigging of *Redoutable.* But the day was a total and crushing victory for the British. *Victory* was towed to Gibraltar where emergency repairs were made before it returned to England with Nelson's body.

Paid off

After a refit at Chatham, *Victory* returned to service as flagship of the Baltic Fleet, engaged in blockading Russia and protecting convoys of naval supplies from Sweden, and also acted as escort to a military convoy to Spain. In 1812 the ship was paid off. Its role in two major victories, and its association with Nelson saved it from the fate of most ships. From 1824 it was designated as flagship of the commander-in-chief at Portsmouth, and it remains in commission as flagship of the Second Sea Lord (C-in-C Naval Home Command).

Living museum

Considerable repair and renewal work was required on *Victory*'s return from Trafalgar, after which it remained active until 1812. For a time it was used as a depot ship, and housed a Naval School of Telegraphy from 1889 to 1904. *Victory* remained afloat, though in increasingly poor condition, until on 12 January 1922 it was moved into the No. 2 dry dock at Portsmouth for permanent keeping. A German bomb did slight damage in 1941. In December 2011 a nine-year restoration programme began. The ship remains an authentic survival of the greatest days of the sailing man of war.

Santísima Trinidad (1769)

The largest warship of the eighteenth century, with four decks of guns, the Spanish flagship was engaged in two of history's great naval battles, at Cape St Vincent and Trafalgar. It was known as the 'Escorial (royal palace) of the Seas'.

The Royal Shipyards at Havana, Cuba (then a Spanish possession) were a major building site for warships. Costs were less than in Spain and there were large timber resources, especially of hardwoods not available in Spain, like the American cedarwood used in *Santísima Trinidad*. It was the seventh Spanish warship of the name, confirmed by a royal order on 12 March 1768.

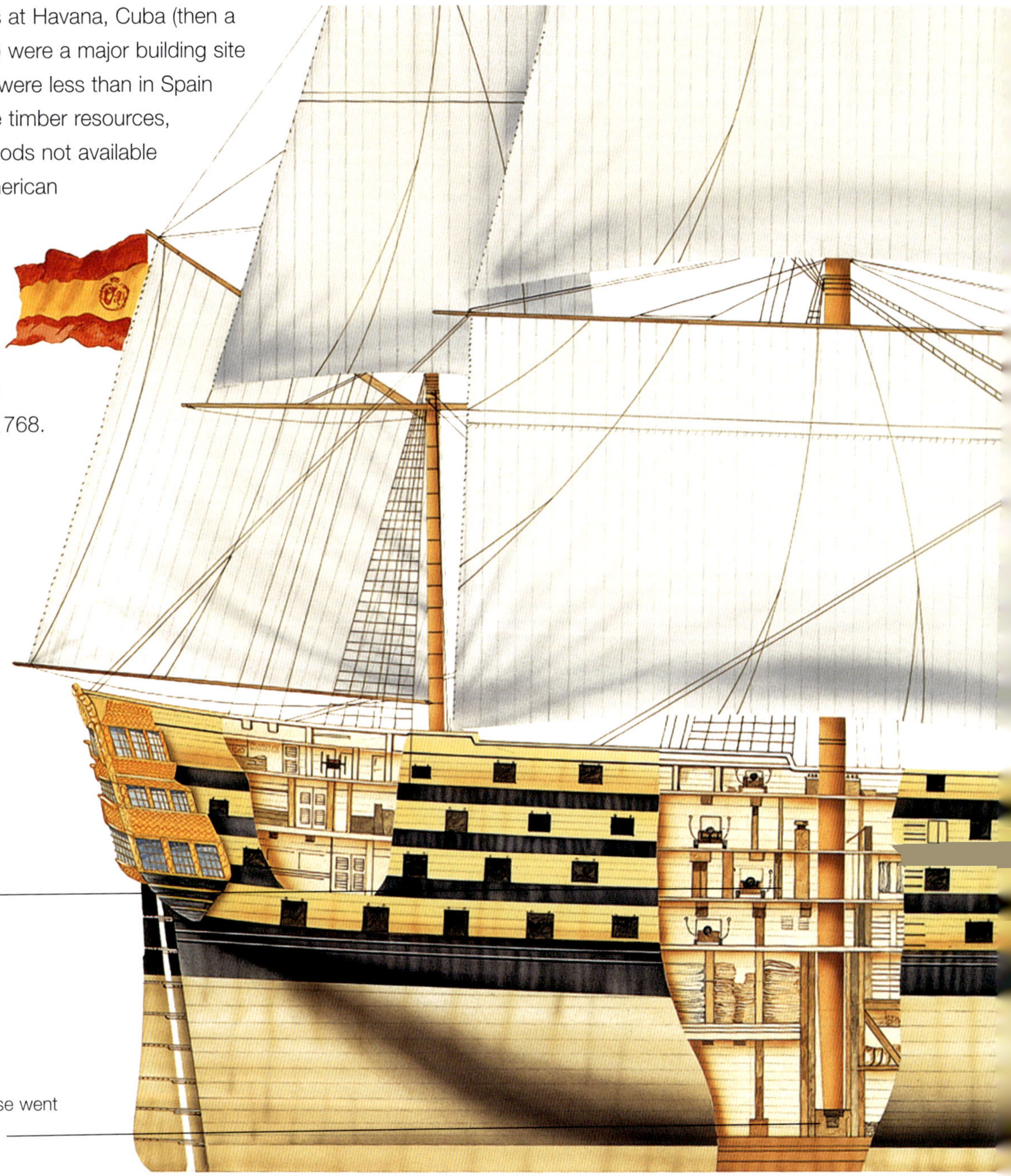

Riding bitts
The anchor cable was hauled in and paid out over these massive wooden bitts.

Pumps
Hand-worked, these went down to keel level.

Epitome of the first-rate 'ship of the line', *Santísima Trinidad* was designed to lead a fleet into battle and to withstand a heavy cannonade. The concept of staying-power in the face of gunfire was becoming increasingly important.

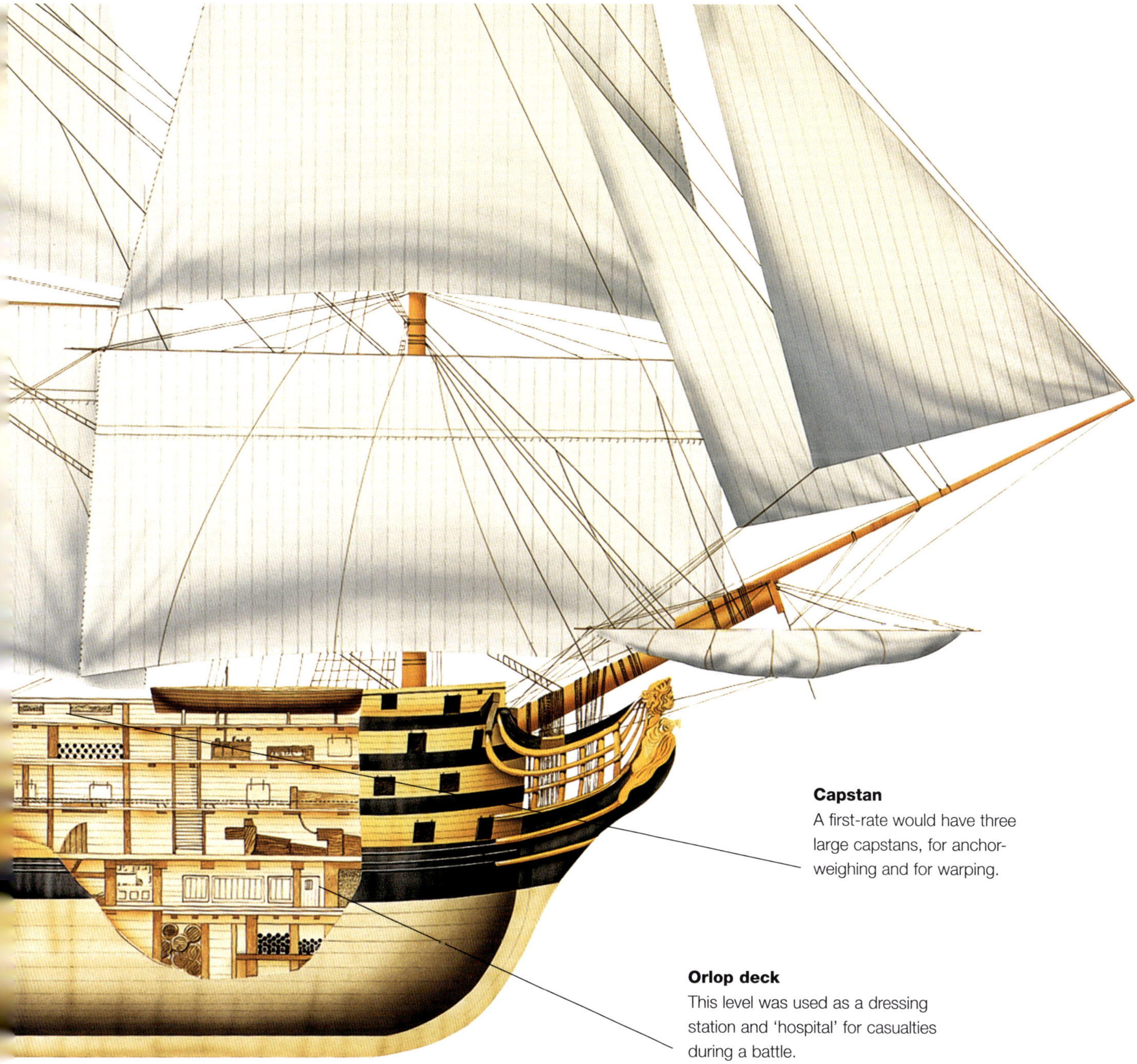

Capstan
A first-rate would have three large capstans, for anchor-weighing and for warping.

Orlop deck
This level was used as a dressing station and 'hospital' for casualties during a battle.

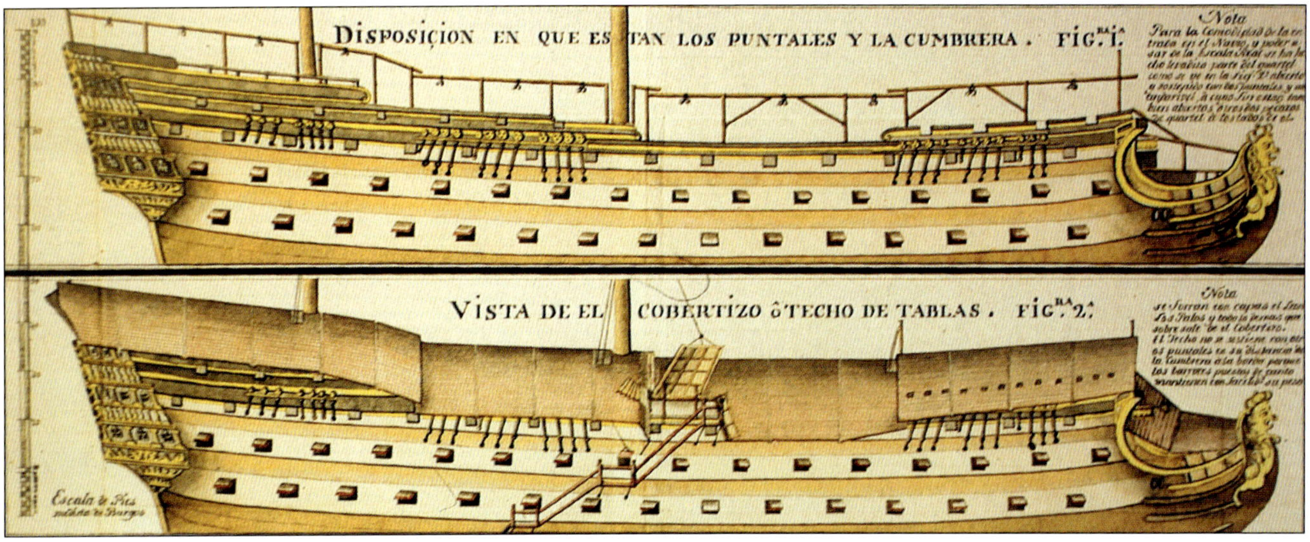

These eighteenth century scale drawings are guides to the installation of the supports for a canvas roofing to cover the entire upper deck. They were made before the vessel's conversion to a four-decker.

Its designer was the King's naval architect Matthew or Mateo Mullan, an Irishman, and building was supervised by his son Ignacio.

It was launched as a three-decker of unusually large dimensions. Spanish shipbuilding was of high quality, perhaps the best of any nation. The ships were strongly built and generally of larger size for their gun-rating than British vessels, which made them both more stable as gun-platforms and better able to withstand attack.

A Spanish 70-gun ship was about 1540 tonnes (1700 tons) compared to the 1134 tonnes (1250 tons) of a comparable British ship. This tradition of size and strength gave the builders of 1769 confidence to construct the largest warship of the time. Ships of this size were rarities: between 1750 and 1790 the British Navy had only six ships of 100 guns. The French also built a few very large ships. In 1788 the French *Commerce de Marseille* exceeded *Santísima Trinidad* in size, being 63.5m (208ft 4in) long, with a beam of 16.6m (54ft 9in), but carried fewer guns, 118 on three decks (captured by the British in 1793, it was broken up in 1802), and *Océan* and *Orient*, of 1790 and 1791, carried 120 guns.

Years of action

In its first years the ship was probably not in commission but held in readiness. With the declaration of war by Spain on Great Britain in July 1779, it entered service as flagship of the Spanish fleet, under Admiral Luis de Cordoba y Cordoba, operating with allied French ships in the English Channel and the western approaches. In August 1780 it led an action which resulted in the capture of 55 British merchant vessels from a convoy. In 1782 it participated in the second siege of Gibraltar, as flagship of a combined fleet 48-strong of Spanish and French ships, but failed to intercept a British relief convoy.

Increased firepower

In 1795, in a bold enhancement of its gun-power, a fourth deck was installed, joining the forecastle to the quarter-deck and raising the number of cannon carried from 112 to 136. This made *Santísima Trinidad* by some way the most heavily-armed ship of its time. Back in service in 1797, it

Specification (1768)

Dimensions	Length 60.1m (200ft), Beam 19.2m (62ft 9in), Draught 8.02m (26ft 4in) Displacement c4309 tonnes (c4750 tons)
Rig	3 masts, square-rigged
Armament	(1768) 30 36-pounder, 32 24-pounder, 32 12-pounder, 18 8-pounder guns
Complement	950

was the Spanish flagship at the Battle of Cape St Vincent on 14 February, and suffered major damage, partially dismasted and with over half the crew killed or wounded. *Santísima Trinidad* struck her colours to HMS *Orion*, but before the British could take possession, they were signalled away, and the ship was rescued by *Pelayo* and *Principe de Asturias*, and limped back to Cadiz for repair.

Particularly after the construction of the fourth deck, giving the ship a very high freeboard exposed to sidewinds, *Santísima Trinidad* did not have good sailing qualities and gained the nickname 'El Ponderoso'. Unlike contemporary French and British naval ships, its hull was not copper-sheathed. A further disadvantage, according to French observers, was a poorly-trained crew and the poor quality of many of the guns. With the greater part of the Spanish fleet, the ship's home base was Cadiz.

In the course of its 38-year plus career, the *Santísima Trinidad* was careened or refitted three times, and spent almost 20 of those years out of service. This last was typical of ships in other navies: if there was no war on, crews were discharged and the ship held 'in ordinary'. Ships in reserve had their guns removed, to reduce strain on the innumerable joints and brackets of the hull and gun-decks.

Surrender at Trafalgar

At Trafalgar, captained by Francisco Javier Uriarte and carrying the pendant of Rear Admiral Baltasar de Cisneros, it was flagship of the Spanish squadron, painted dark red with white stripes. In line just ahead of Admiral Villeneuve's *Bucentaure*, it was in the thick of the central battle, heavily raked by broadsides from HMS *Neptune*.

After four hours, by 2:12pm, all three masts were gone; an eyewitness wrote: 'This tremendous fabric gave a deep roll, with a swell to leeward, then back to windward, and on her return every mast went by the board, leaving it an unmanageable hulk on the water.' The ship was compelled to surrender (as painted below in the *Surrender of the Santísima Trinidad to Neptune, The Battle of Trafalgar, 3 PM, 21st October 1805* by Lieutenant Robert Strickland Thomas).

After the battle it was taken in tow by HMS Prince, but in the storm which followed, the tow could not be held, and *Santísima Trinidad* was scuttled on 22 October.

Santa Ana (1784)

A first-rate of 112 guns, *Santa Ana* was the first of a class of eight ships intended to provide the central strength of the Spanish Navy. It was Spain's flagship at Trafalgar on 21 October 1805.

Naval architecture was a well-established science by the early eighteenth century. By mid-century Spanish shipbuilders were applying lessons learned from English and French designers to their own established techniques, particularly relating to wood treatment and construction methods. Jose Romero y Fernandez de Landa, *Santa Ana*'s designer, was a scientific builder, author of a textbook on the construction of warships, published in 1784.

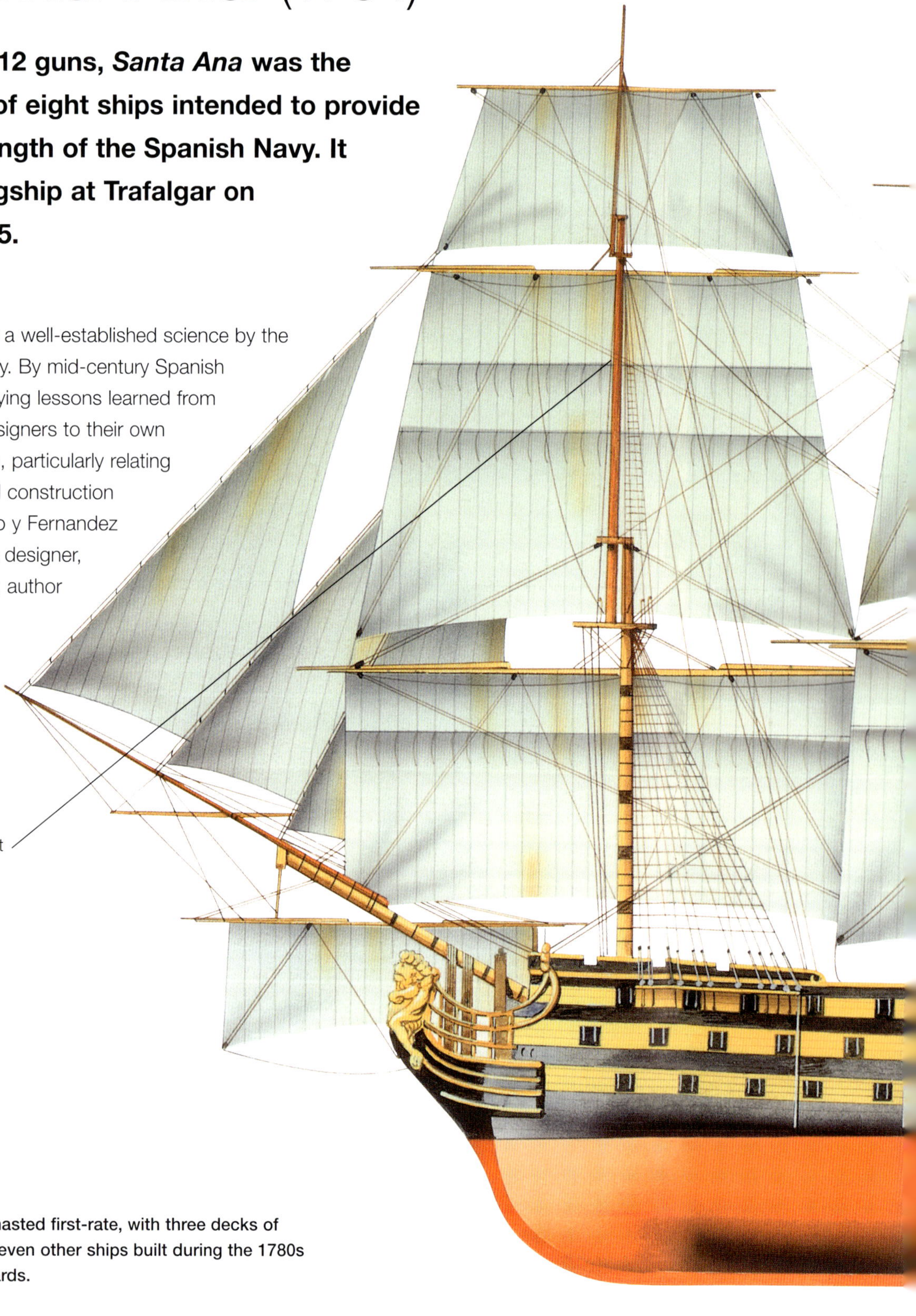

Masts
With stays and rigging shot away, the masts were vulnerable. *Santa Ana* lost all three at Trafalgar.

Santa Ana **was a three-masted first-rate, with three decks of guns: the prototype for seven other ships built during the 1780s at Spanish and Cuban yards.**

Yards
Note the footropes, giving a foothold to the sailors who had to man the yards in order to furl the sails.
Sails
In addition to the standard square-rig arrangement seen here, studding sails would be rigged at the outer edges of the main sails, and staysails on all three masts.
Hull
Like French ships, the Spanish ships of the line had a high freeboard, with the lowest gun deck – and heaviest cannon – well above the waterline.

Specification

Dimensions	Length 56.14m (184ft 2in), Beam 15.5m (50ft), Draught 7.37m (23ft), Displacement 2543 tonnes (2803 tons)
Rig	3 masts, full-rigged ship
Armament	(1805) 30 36-pounder, 32 24-pounder, 10 8-pounder cannon; 10 48-pounder, 2 32-pounder, 6 24-pounder howitzers; 4 4-pounder swivel guns
Complement	1000-plus

Built at El Ferrol, the vessel was launched on 29 September 1784 and first put to sea on 24 November that same year, though not commissioned until 28 February 1785. It was based at Cadiz and maintenance work was done at the La Carraca Arsenal. In January 1787 it was dry docked at La Carraca, and in June 1791 was careened there and rotten timbers were replaced.

Facts and figures

Santa Ana had seven anchors with a total weight of 20,457kg (45,100lb). Its ballast, in iron (small ball shots, and iron pieces) occupied around 20 per cent of the length, placed amidships and surrounded by stone ballast. An indication of the care and attention put into construction is given by the ballast-laying instructions. First tar was applied to the holding timbers, then a layer of zulaque (sticky clay or cement), 102mm (4in) thick was applied to the floor-heads. Ground brick was added in alternate layers with iron and fine mix to fill up the space up to 305mm (12in) above the floor-heads. Above this, only the mix of gravel and brick was laid to the heads of the first futtocks. The aim was to ensure that the ballast, around 81 tonnes (90 tons), was densely packed and could not shift in any direction with the pitching and rolling of the ship.

Inactivity

In 1794 the ship was given a full careen at La Carraca, rearmed in January 1797 and stationed at Cadiz. It was with the fleet blockaded in Cadiz in February 1798. In September that year it was careened again at La Carraca and copper sheathing was applied to the bottom. On 21 July 1799 it grounded at the Rota naval base but was refloated. A new keel was fitted in the course of 1800, and *Santa Ana* remained on service until 1802, when it was again disarmed. Most of its time until 1805 was spent disarmed either through being laid off or having repairs and maintenance; its periods of activity totalled approximately five years out of twenty-one. Disarming meant the removal of everything not integral to the hull structure, including the lower masts, bowsprit and ballast. Everything else was removed and stored away, including the rudder.

In January 1805 preparations for a new spell of active duty began. Careened in La Carraca in September of that year, it was newly rearmed and re-manned when it put to sea with the Combined Fleet, as flagship of the Spanish second-in-command.

Battle at sea

In the Battle of Trafalgar, *Santa Ana* carried the flag of Vice-Admiral Alava and was captained by Jose Gardoqui. It seems it was painted white with black stripes, though some accounts state it was wholly black. Its position in the Combined Fleet's line brought it against the British fleet's lee division, headed by HMS *Royal Sovereign,* flagship of Vice-Admiral Collingwood, under whose drills the ship's gunnery

The art of careening

Careening was normally done afloat. It was a tricky exercise, with everything moveable packed away or lashed tight. The hull was then tilted until the keel was exposed above the water. This was achieved by hauling on the lasts from floating pontoons or from the shore or jetty. The hull would have internal supports and the masts would be propped to avoid strain. Ballast would be shifted to the tilt-side and extended forward to minimise the need for excessive pull on the foremast. Full careening would involve removal and replacement of the shell planking, but often only partial careening was done, for replacement of damaged or rotted planking. *Santa Ana*'s careening programme was typical, with treatment every three years to remove weed and barnacle growth and check the state of the timbers.

The handsome lines, intricate rigging and good performance of the class have inspired many modellers over the years. The arc-shape design of the transom, often shown painted blue, is a distinctive feature.

was the best in the British fleet. *Royal Sovereign* crossed just abaft of *Santa Ana* and fired a double-shotted broadside into the stern, which put 14 guns out of action and caused many casualties. The two ships were then locked together for a time, with *Royal Sovereign* against *Santa Ana*'s starboard bow, in a devastating cross-fire that continued for almost two hours. 'They fought us pretty tightish', reported a British midshipman. *Santa Ana*'s mizzen topmast was shot away, and after about an hour and a quarter all its three masts had fallen over. At about 14:20 it struck to *Royal Sovereign*.

Two days after the battle, *Santa Ana* was recaptured by a Spanish frigate squadron and towed back to Cadiz. When the French invaded Spain in 1808 it was still under repair and took no part in the Peninsular War. With a sister ship, *Principe de Asturias*, it was moved to Havana in 1810, but saw no further action. It eventually sank at the Havana Arsenal, in 1816.

Bucentaure (1803)

A new, handsome, well-proportioned 80-gun ship of the line, it was the flagship of Admiral Villeneuve, commander-in-chief of the Combined Fleet at the Battle of Trafalgar.

Bucentaure was the first of a class of 80-gun ships for the Imperial French Navy. Sixteen were launched between 1803 and 1815, with a further five up to 1824. The designer was Jacques-Noel Sané. Sané was an advocate of uniformity in ship design, and beginning with *Tonnant* in 1789, was responsible for a series of excellent ships. Armed with 84 guns, they were classified as second-rates, but outgunned the typical British 74-gun ship and were easier to handle than the massive 100-gun first-rates.

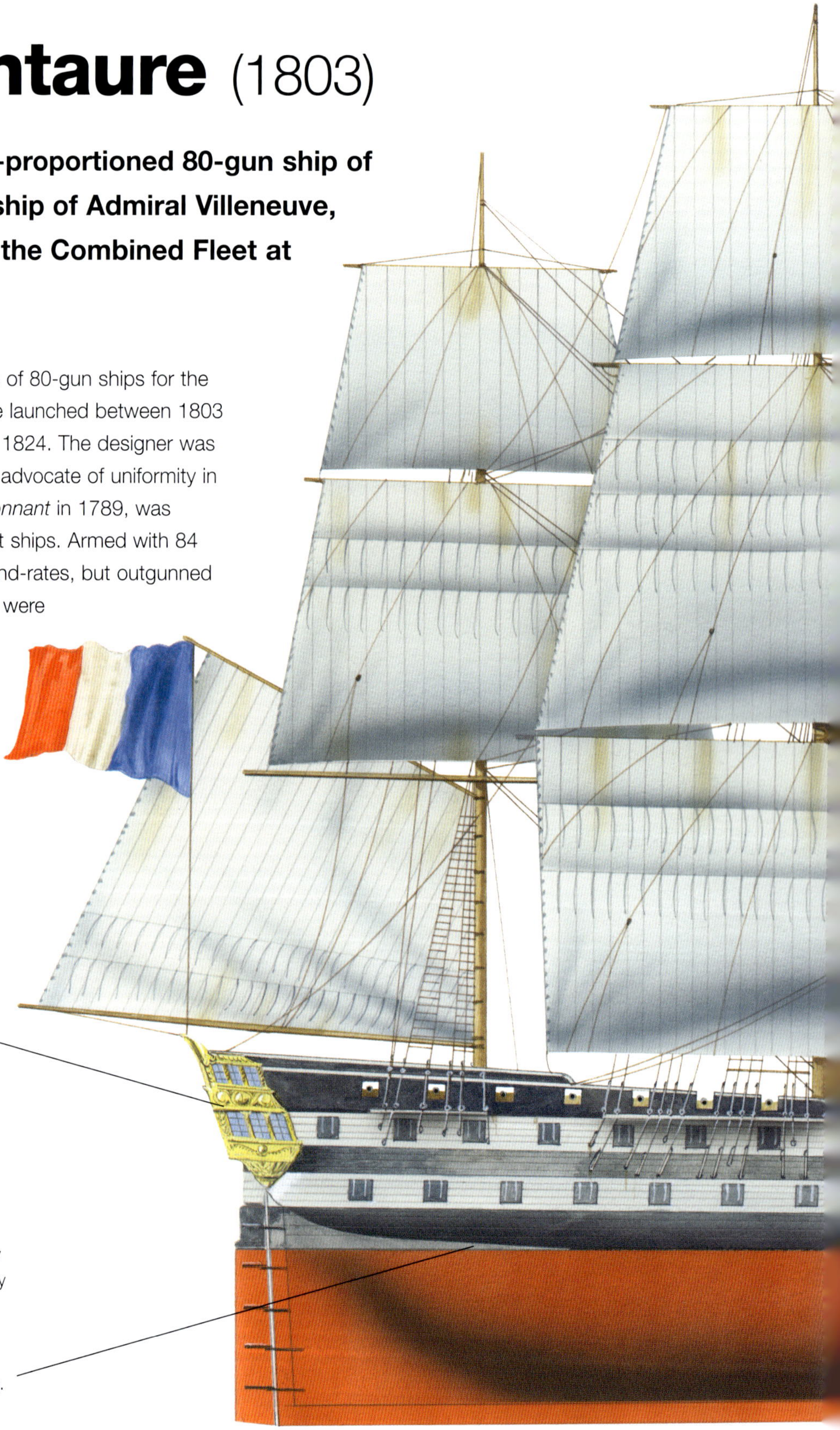

Weak stern
The expectation of exchanging side-on fire left the stern a weak and exposed area. *Bucentaure* was virtually incapacitated by a single devastating treble-shotted broadside through the stern.

Hull lines
The hull lines of French men of war were usually more carefully and subtly designed than was the case with British ships, giving a slight edge in speed and sea-handling.

Bucentaure

Bucentaure's name commemorated the French capture of Venice in May 1797: it was taken from the name of the Doge's state barge, *Bucintoro.*

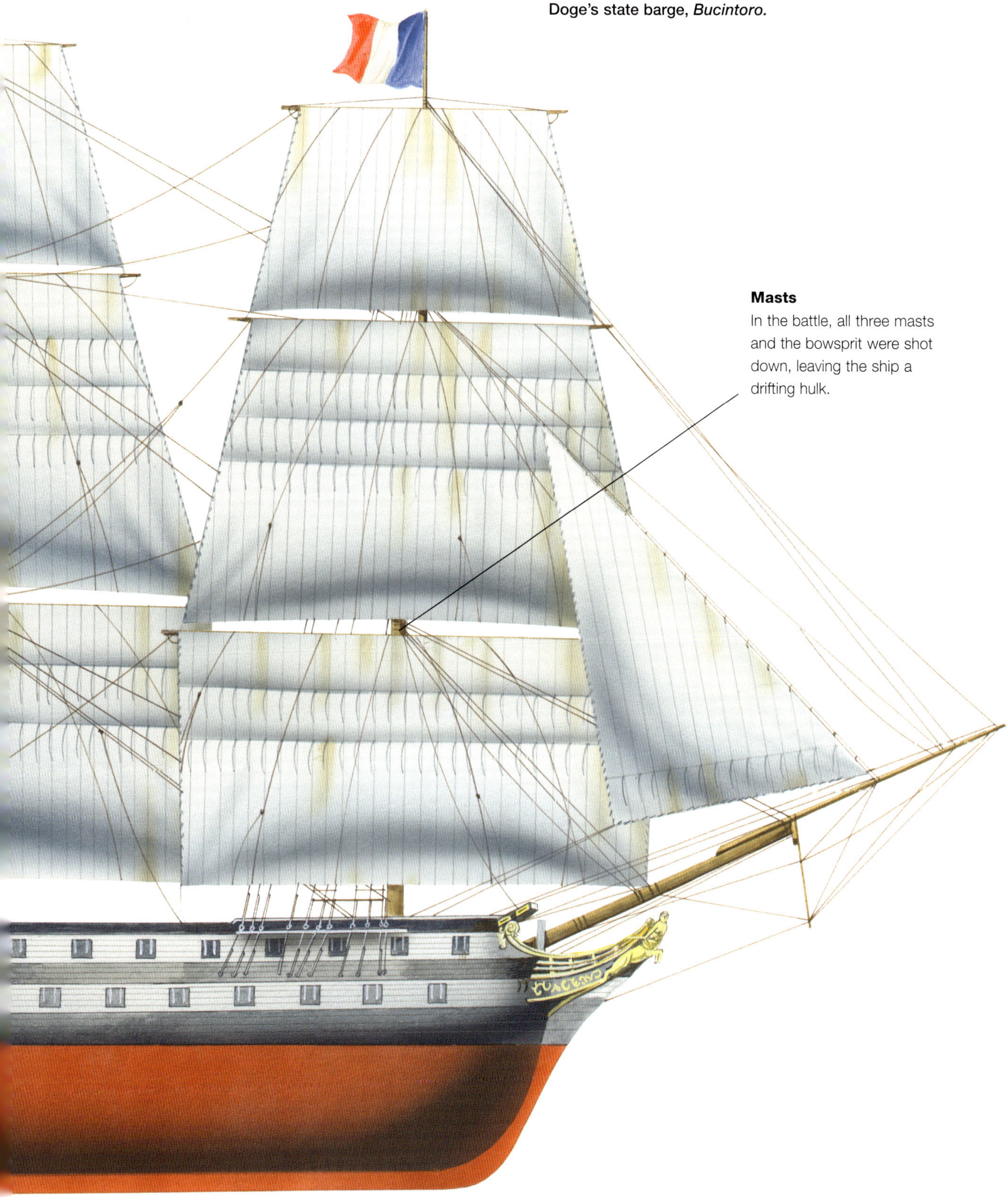

This painting by Auguste Mayer (1805–90) was long thought to show the dismasted *Redoutable* but has been shown in fact to represent *Bucentaure*. The British ship's stern bears the name 'Sandwich' but HMS *Sandwich* was not at Trafalgar.

Longer than their British counterparts, more solidly constructed and heavier, they were well-rigged and can fairly be claimed to be the best ships of the time. Some commentators consider there to be a single class from *Tonnant* to the final example, *Vesuvio*, not launched until 1824 and sold to the Kingdom of the Two Sicilies.

The great problem for the French captains was not the quality of their ships but that of their crews. The British Navy had worked harder to train its officers and crews, helped in this respect by having far more ships at sea, and for longer periods. The British crews were healthier, with the shipboard scourge of scurvy largely prevented. The French captains considered their crews to be a rabble. There is a phrase attributed to Nelson: 'The best navy in the world would be made of French ships and English crews.' But English crews also contained their full share of pressed men, freed prisoners and army deserters.

Launch and service

Laid down at Toulon in November 1802, *Bucentaure* was launched on 13 July 1803 and commissioned in January 1804. From first service it was a flagship, at first of Vice Admiral Latouche-Tréville, who died on board, to be succeeded by Vice Admiral Pierre de Villeneuve from 6 November 1804. In September 1805, the Combined Fleet was at Cadiz, when an order from Napoleon instructed Villeneuve to embark troops and set sail for an invasion of Naples. Cadiz was already closely watched by the British, and there was little chance of leaving it without a battle. But Villeneuve learned that he was to be replaced by Admiral Rosily, and despite having decided in September that the Combined Fleet was not capable of action, he resolved to take his ships to sea and vindicate his post in the eyes of Emperor Napoleon.

They left Cadiz on 21 October. On paper, Villeneuve had a superior force, with 33 ships of the line (18 of them French) and seven frigates, and with 2856 cannon at his disposal, while Nelson had 2314, on 27 ships of the line and six frigates. But the British admiral had seven three-deckers while Villeneuve had only four.

Into battle

Bucentaure was commanded by Captain Jean-Jacques Magendie, and in the line it took a central place. As the British, in their two divisions, advanced, Villeneuve's last signal was similar to Nelson's: 'Every ship which is not in action is not at its post, and must take station to bring itself as speedily as possible under fire.' *Victory* fired a broadside that ripped into the transom of *Bucentaure* at a range of only 9m (30ft), sweeping the decks with shattering effect. In a few minutes *Victory*, *Bucentaure*, *L'Indomptable* and HMS *Temeraire* were all abreast of or inboard each other, rolling together, spars crashing, gunshots blasting off. At 13:40, shots from HMS *Conqueror* brought down the main and mizzen masts.

'A mass of wreckage'

At 13:45, now drifting helplessly in the midst of the battle, with the bowsprit and all three masts fallen, and half the crew killed or wounded, *Bucentaure* was described as 'a mass of wreckage' by the Captain. Villeneuve tried to have his barge launched, to transfer his flag to a ship still able to fight, but the boat was crushed beneath fallen spars. With no alternative but to surrender *Bucentaure* struck its colours to *Conqueror*. Villeneuve and Magendie were taken prisoner. A British prize crew was put on board, with the surviving French crew held on board as prisoners, and the ship was attached by a towline to *Conqueror*, but the line parted. The Frenchmen managed to break out and retake the ship, but in the storm which arose on the 23rd, it was unmanageable. The ship struck a reef off Cadiz Bay, and foundered, with a handful of survivors.

Specification

Dimensions	Length 59.3m (194ft 6in), Beam 15.3m (50ft 3in), Draught 7.8m (25ft 6in), Displacement 1455 tonnes (1604 tons)
Rig	3 masts, full ship rig
Armament	30 36-pounders, 32 24-pounders, 18 12-pounder cannon; 6 36-pounder howitzers
Complement	866 sailors and marines

The crew

The crew was divided into gentlemen-officers, petty-officers (illustrated right), men, and supernumeraries. The first category numbered 14, including a Commander *(capitaine de frégate)* as No. 2 to the Captain *(capitaine du vaisseau)*; purser and a chief medical officer, plus nine midshipmen. The second category included two boatswains, and 25 boatswains' mates; two master-helmsmen or quartermasters and 13 mates; 55 master gunners and their mates; two master blacksmiths and 13 mates; a master-carpenter and seven mates; a master caulker and seven mates; a master-sailmaker and two mates; and a master-at-arms. There were four rated grades of seaman, each with 95 men, 125 unrated new recruits, 60 powder-boys and 129 marines. Supernumeraries were three armourers, five surgeons, nine cooks and 11 servants.

Gloire (1858)

The first seagoing armoured warship, it initiated something of a battleship race between Great Britain and the France of Napoleon III in the mid nineteenth century.

Steam power had been an accepted aspect of naval ships since around 1830, but initially its use was confined to paddle-wheeled tugs. Paddle-wheels on a man of war had obvious disadvantages, though a number of paddle frigates were built in the 1840s. Development speeded up in the 1850s. French strategic planners realised that steam and iron had effectively cancelled out the British Navy's great superiority in numbers, and that both nations had to build new fleets. Though they shared a starting-point, it was not quite a level base as the British iron and

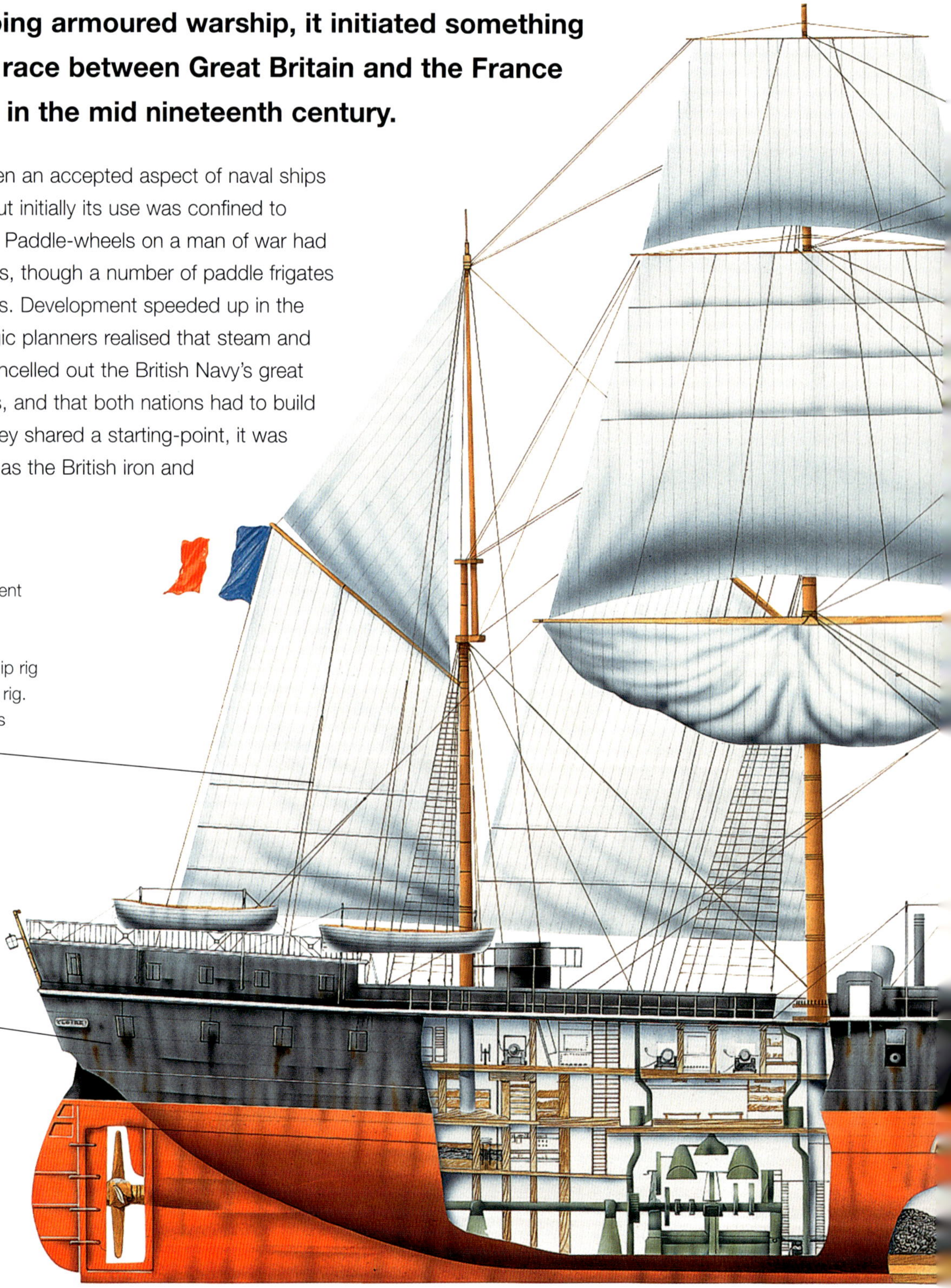

Rigging
Gloire was rigged in different ways at different times. Originally rigged as a barquentine, later a full ship rig was applied, then barque rig. The final arrangement was fore and aft sail only.

Wood v iron
Three of France's first four ironclads, all ordered in March 1858, were wooden-hulled. This was due to France's lack of ironfounding capacity at this time. Only *Couronne* had a completely iron hull.

Armour-plating on warships was the inevitable result of heavier guns, the development of explosive shells, and technical developments in metallurgy.

Brute force
Though not designed for ramming, *Gloire*'s 119mm (4.7in) iron-plated bow would have stove in the side of any wooden ship.

Storage
The engines and coal bunkers needed a large amount of space, quite apart from the storage requirements of a man of war. Steam warships had to be larger, and of course were more expensive.

Stokehold
The stokeholds of steamers were always highly unpleasant places in which to do hard physical work, especially before forced-draught ventilation was introduced.

Gloire

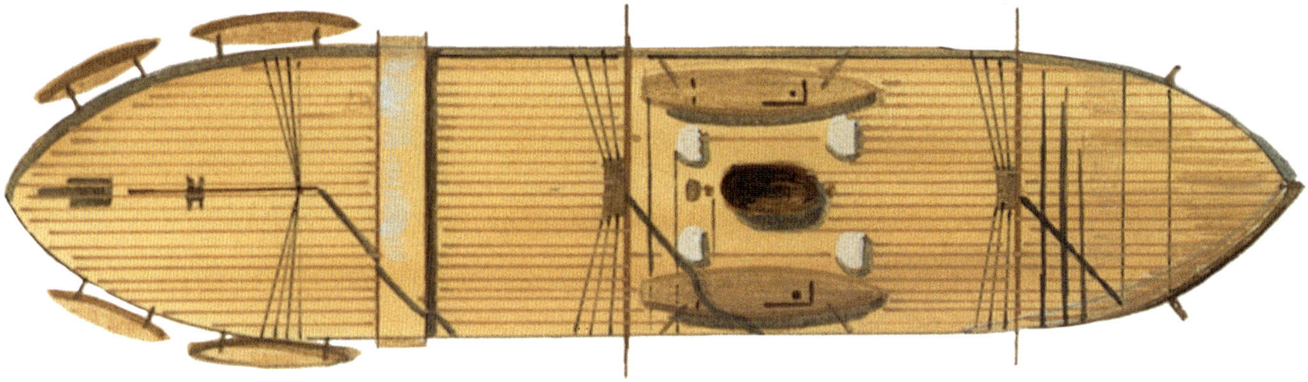

The weather deck is almost entirely clear, apart from funnel, ventilators, and a bridge structure (with a form of conning tower) just before the mizzen mast. All guns are on the main deck below.

mechanical engineering industries had not yet lost the head start they had acquired earlier in the century.

Although the British and French had been allies against the Russians in the Crimean War (1854–56) there was still considerable rivalry between them. One of the innovations of the war had been the French 'floating batteries', flat-bottomed ironclad hulls with limited capacity for self-propulsion by sail or low-powered steam engine, which had been towed to the war zone and used to bombard shore targets. Their success prompted the design of more seaworthy craft of similar type.

Age of iron

By 1858 both the French and the British had built or converted 32 wooden-hulled ships of the line as steamships, and both had more on the stocks: 6 in Britain, 10 in France. Another new development, of rifled guns firing shells at high velocity, convinced French naval planners that iron armour was now essential and the order was given to cease construction of wooden hulls in 1857. In that year Stanislas Dupuy de Lôme was appointed as head of naval construction. A brilliant and innovatory naval engineer, he had already designed the first steam screw-driven warship, *Napoléon*, launched in 1850, and *Gloire* was designed on broadly similar lines. It was not the first ironclad but it was the first true seagoing warship of the kind.

The hull of this new ship was of wood, protected by 119mm (4.7in) iron plates, and with a single iron deck carrying 36 162mm (6.4in) muzzle-loading rifled (MLR) guns, each weighing 4.5 tonnes (5 tons). These guns fired explosive shells, intended to pierce a wooden hull and then explode.

Gloire was laid down at Toulon Naval Yard in May 1858, launched 1859, and commissioned in July 1860. In appearance it was compact, even squat, lying low in the water, with a straight bow and pointed stern. The Naval Committee of Works had specified that new armoured ships should manage a speed of 13 knots, have armour of at least 100mm (4in) thickness to a depth of 1.3–1.7m (4–5ft) below the waterline, a stern shaped to give protection to the screw, and a full ship rig for cruising without resort to steam power.

Gloire fulfilled these requirements, though its speed in service was probably a little under 12 knots. The thick hull, without openings apart from the gun-ports, made the

Specification

Dimensions	Length 77.9m (225ft 6in), Beam 17m (55ft 9in), Draught 8.5m (27ft 10in), Displacement 5095 tonnes (5617 tons)
Propulsion	8 boilers, 2-cycle trunk steam engine, 1864kW (2500hp), single screw; 3 masts, full ship rig
Armament	(1867): 6 244mm (9.6in) MLR guns; 2 198mm (7.8in) MLR guns
Armour	Belt 119mm (4.7in)
Speed	12.5 knots
Complement	550

interior stuffy and smoky and often hot, and it was not a comfortable ship to serve on. It served for nine years before receiving a comprehensive overhaul and refit, which included replacement of its original armament by fewer but more up-to-date heavier guns (see Specification).

French viewpoint

The French Navy took a different view of steam power to the British Admiralty. While the British still saw steam as an auxiliary to sail, the French, without discarding sail as yet, saw the engine as the prime mover, and the propellers on French ironclads were not dismountable. This partly reflects the shorter strategic range of the French fleet, but also shows a greater willingness to embrace the new, smoky, clanking technology and make it as effective as possible. *Gloire* carried 1100m^2 (11,840sq ft) of sail area.

Gloire's low position in the water, with its gun deck only 1.9m (6ft 3in) above the waterline, reminiscent of the Crimean floating batteries (which would be used again, against Italy in the Adriatic in 1859), restricted its usefulness as an ocean-going ship. But in any case, construction of the British HMS *Warrior* two years after *Gloire* made the French ship effectively obsolescent. In 1879 *Gloire* was removed from the Navy list and in 1881 it was broken up for scrap.

Armour plating

The hull plating was 109mm (4.3in) thick at the upper deck level, thickening to 119mm (4.7in) below, and the wooden hull itself was 660mm (26in) thick at the maximum. Iron plating was not as effective as copper at deterring the growth of plant-life on the underwater hull and *Gloire* was expensively fitted with copper plates in the 1860s. But the interaction of copper, iron and seawater corroded the iron and destroyed the copper's deterrent effect on weeds and barnacles. A layer of wood had to be interplaced in order to prevent electrolytic interaction between the metals. At the time of construction, the armour would have defeated any ship-borne artillery, but the lead of armour over shell-power was brief. The naval shell gun, invented by Joseph Paixhans, introduced by the French Navy in 1841, and swiftly adopted by other navies, was being improved. A game of 'leap-frog' between better armour and more effective guns would continue into the twentieth century.

Gloire and *Donawerth,* by the marine artist Louis le Breton (1818–1866). *Donawerth* was a 90-gun ship of the line launched in 1854, and twice renamed, *Jean Bart* in 1868 and *Cyclope* in 1886.

Victoria (1859)

With its sister ship *Howe, Victoria* was the last wooden-hulled three-decker to be built for the British Navy. Already obsolescent despite its steam propulsion, it had an active life of less than 10 years.

It was an era of distinct change. The last sail-driven ships of the line were ordered for the Royal Navy as late as 1848 (the 80-gun *Orion* class). Lord Auckland, First Lord of the Admiralty 1846–49, had defined the new requirement for design: '…the manner in which the screw auxiliary may be best combined with good sailing qualities.'

The Crimean War of 1854–56 was a long-distance war for France and even more so for Britain, requiring transport through the Mediterranean and into the Black Sea. Steam propulsion, with its speed and independence of the wind, really came into its own. But the Royal Navy still had relatively few steam-powered warships and a hurried programme of conversion began. *Victoria* however was planned from the start as a steamship. In the British Admiralty's

Mainmast
The machinery arrangement meant that the mainmast had to be stepped on the orlop deck rather than reach to the keel.

Stern walkway
The propeller was fixed, enabling the stern walkway, always popular with officers, to be reinstated.

The two-funnelled *Victoria* combined up-to-date and traditional features. Despite powerful engines it carried a full spread of sail. Note also the stern galleries.

Funnels
Victoria and *Howe* were the only RN ships with funnels fore and aft of the mainmast. Both funnels could be lowered when the ship was under sail.

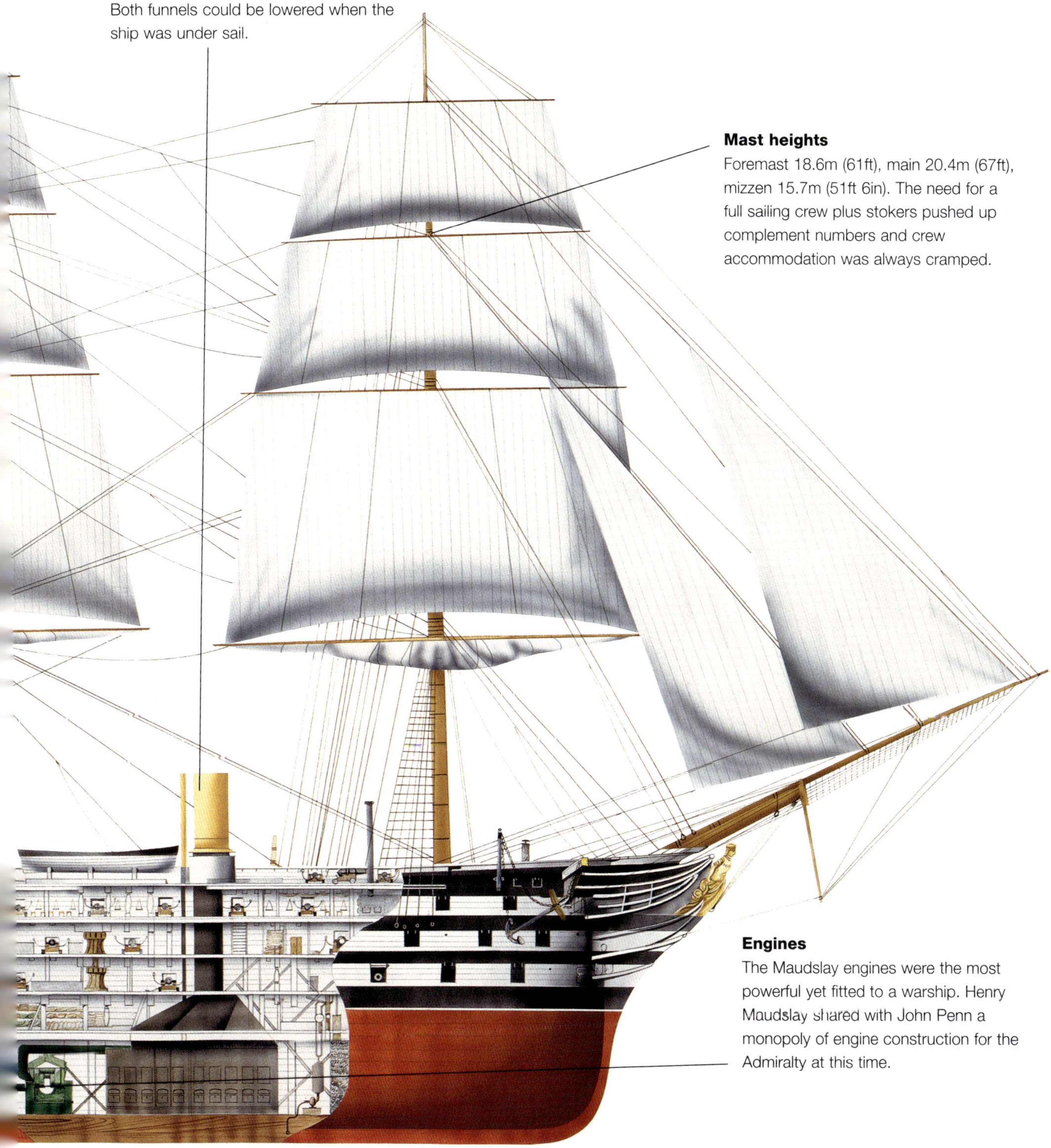

Mast heights
Foremast 18.6m (61ft), main 20.4m (67ft), mizzen 15.7m (51ft 6in). The need for a full sailing crew plus stokers pushed up complement numbers and crew accommodation was always cramped.

Engines
The Maudslay engines were the most powerful yet fitted to a warship. Henry Maudslay shared with John Penn a monopoly of engine construction for the Admiralty at this time.

typical knee-jerk response to French developments, it was intended to rival the 130-gun French three-decker *Bretagne*, which had been laid down as a sailing ship but was converted to steam propulsion while building, and launched in February 1855.

Hybrid

Victoria was laid down at Portsmouth on 1 April 1856 and launched on 12 November 1859. It was a hybrid, a traditional wooden-hulled three-decker of 121 guns with steam engines, and cost £150,578. Eight boilers were arranged laterally in four pairs with the engine between, and it was the Royal Navy's first two-funnelled warship. The hull was heavily strapped on the inner side with diagonal iron riders, 127mm (5in) wide and 25mm (1in) thick, to hold the planking together against the vibrations from machinery and screw. Even so, the seams tended to

Victoria was named by Queen Victoria in person. The overall building shed, unusual at that time, allowed all-weather construction rather than concealment of what was being built.

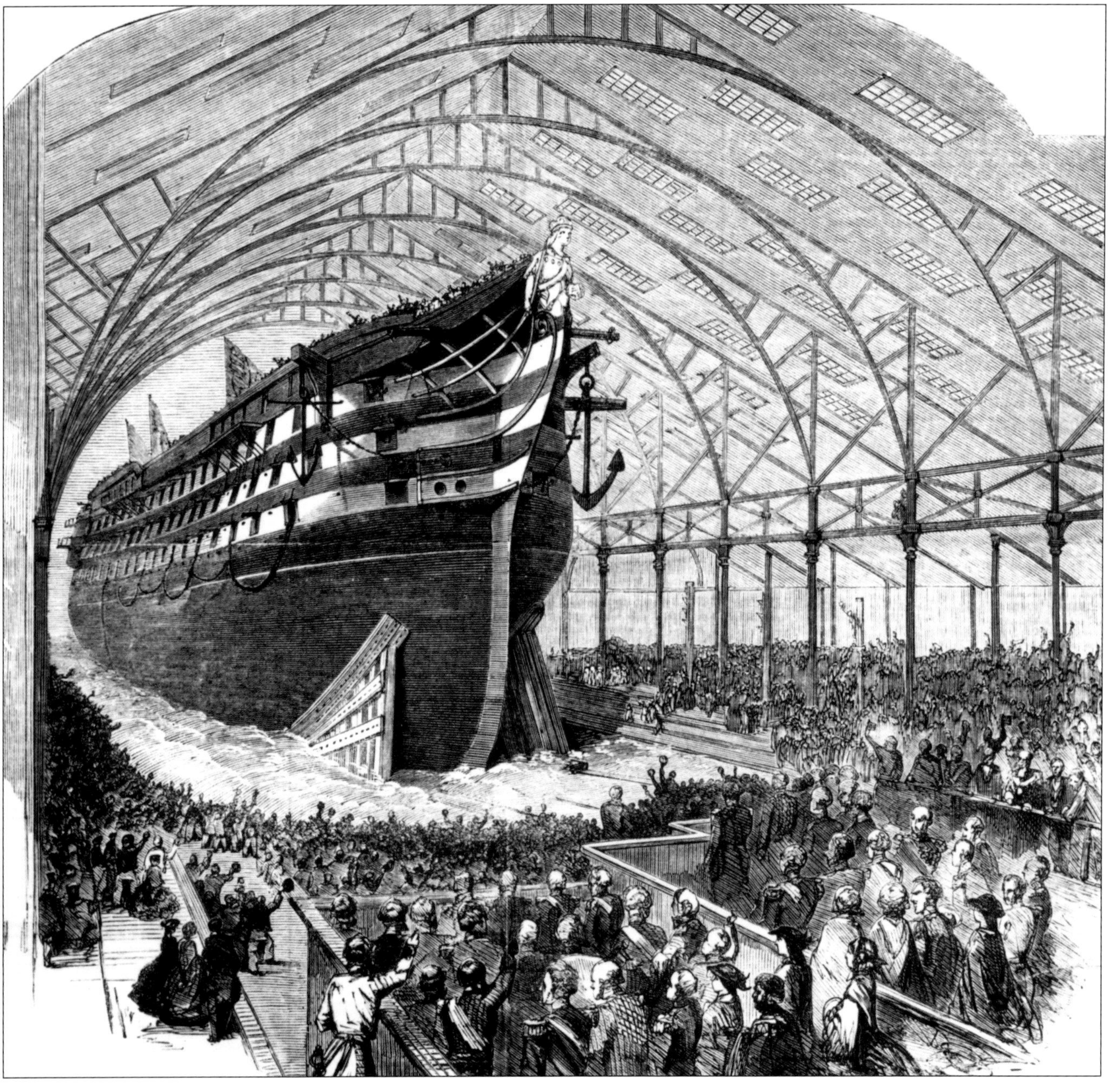

Specification

Dimensions	Length 79.2m (260ft), Beam 18.3 (60ft), Draught maximum 7.8m (25ft 9in), Displacement 6313 tonnes (6959 tons)
Propulsion	Maudslay 3283kW (4403hp), single screw; 3 masts, full ship rig
Armament	62 200mm (8in), 32 30-pounder, 36 32-pounder, 1 68-pounder
Speed	11.79 knots on trials
Complement	1000

separate and it was a leaky ship. Quite apart from the superior shell-resistance of iron, the problems arising from powerful machinery and heavyweight guns in a wooden hull encouraged the use of iron in construction. *Victoria* was the largest wooden-hulled battleship ever built, and was briefly absolutely the largest, until the arrival of HMS *Warrior*. The early engines were bulky, and the weight of machinery, low down in the hull, enabled builders to greatly reduce the amount of ballast carried, or even dispense with it altogether, as in the 101-gun HMS *Conqueror* (1855). But captains had to learn how to trim their ships as the coal bunkers gradually emptied. *Victoria* and *Howe* were both very fast ships for their time; on trials *Victoria* made 11.79 knots and *Howe* achieved 13.56 knots though neither was carrying anything like a full load.

Destructive power

Victoria has been described as 'the acme of the three-decker', at twice the tonnage of HMS *Victory*, and with a far higher destructive capacity. A 68-pounder chase gun was mounted in the bow to give forward fire, and the 200mm (8in) and 30-pounder guns were fitted with sights to enable accurate fire at ranges of at least 1097m (1200yd). These guns fired explosive shells which penetrated wood planking and then burst.

Better sighting and longer range also made greater dispersion of fire possible, so that Victoria could direct fire at more than one target at a time. As a result of this, tactics for naval battles were being revised, and within two years the giant three-deckers were acknowledged as obsolescent, when the House of Commons voted in April 1861 to end the construction of wooden-hulled capital ships. This did not lead to the immediate abandonment of the wooden steam-powered men of war, but major refits stopped, and by the end of the 1860s they were out of use.

The end of *Victoria*

From 1860 to 1864, *Victoria* was held in reserve at Portsmouth, then from 1864, with the number of guns reduced to 102, was flagship of the Mediterranean Fleet, with its main stations at Gibraltar and Malta. The last time the great wooden-hulled ships went on parade was at the Spithead Naval Review of July 1867, before Queen Victoria and the Ottoman Emperor.

In August 1867 *Victoria* was paid off, and technically returned to reserve status, but in fact the old wooden vessels were either converted to other uses or left to rot away. Another HMS *Victoria* was launched in 1887, claimed as the most powerful ironclad in any navy, by which time the old *Victoria* was disarmed. The hull was sold for scrap in May 1893. A month later the new HMS *Victoria* was sunk in a collision with HMS *Camperdown* during drill manoeuvres.

Armament

Victoria's 200mm (8in) guns fired explosive shells. Although these had been in use for some 20 years, they had not proved to be as devastating a weapon as their inventor Joseph Paixhans had supposed, often failing to detonate due to faulty fuses. Shell-fire was more effective in theory and in target practice at this time than in combat. Perhaps more dangerous for a wooden-hulled ship was red-hot cannon balls. These were frequently used by Russian shore forts in the Crimean War and, and from 28 February 1854, the Royal Navy's steam-powered ships were equipped with furnaces and handling gear for firing red-hot shot. Of course this required very careful handling, and only 32-pounder guns of the most robust construction were considered really safe to fire them. Until 1860 British naval guns were, as they had always been, of smooth-bore muzzle-loading type, but it was evident that these were, or very soon would be, outdated. The problem was what to replace them with.

Warrior (1861)

The steam ironclad *Warrior* was planned as a direct response to France's *Gloire.* Laid down at Mare's yard at Blackwall on the Thames on 25 May 1859, launched on 29 December 1860, commissioned at Portsmouth in August 1861 and finally completed on 24 October that year, it cost £377,292.

Its hull construction marked a clear break with the old wooden-hulled tradition. It was to carry 40 large guns on a single deck. But the design showed compromises between the new and the old: *Warrior*'s heavy 'knee bow' design was a convention rather than a structural requirement, and weighed the vessel down at the bow end until a shelter-deck was erected at the poop. The wide stern was

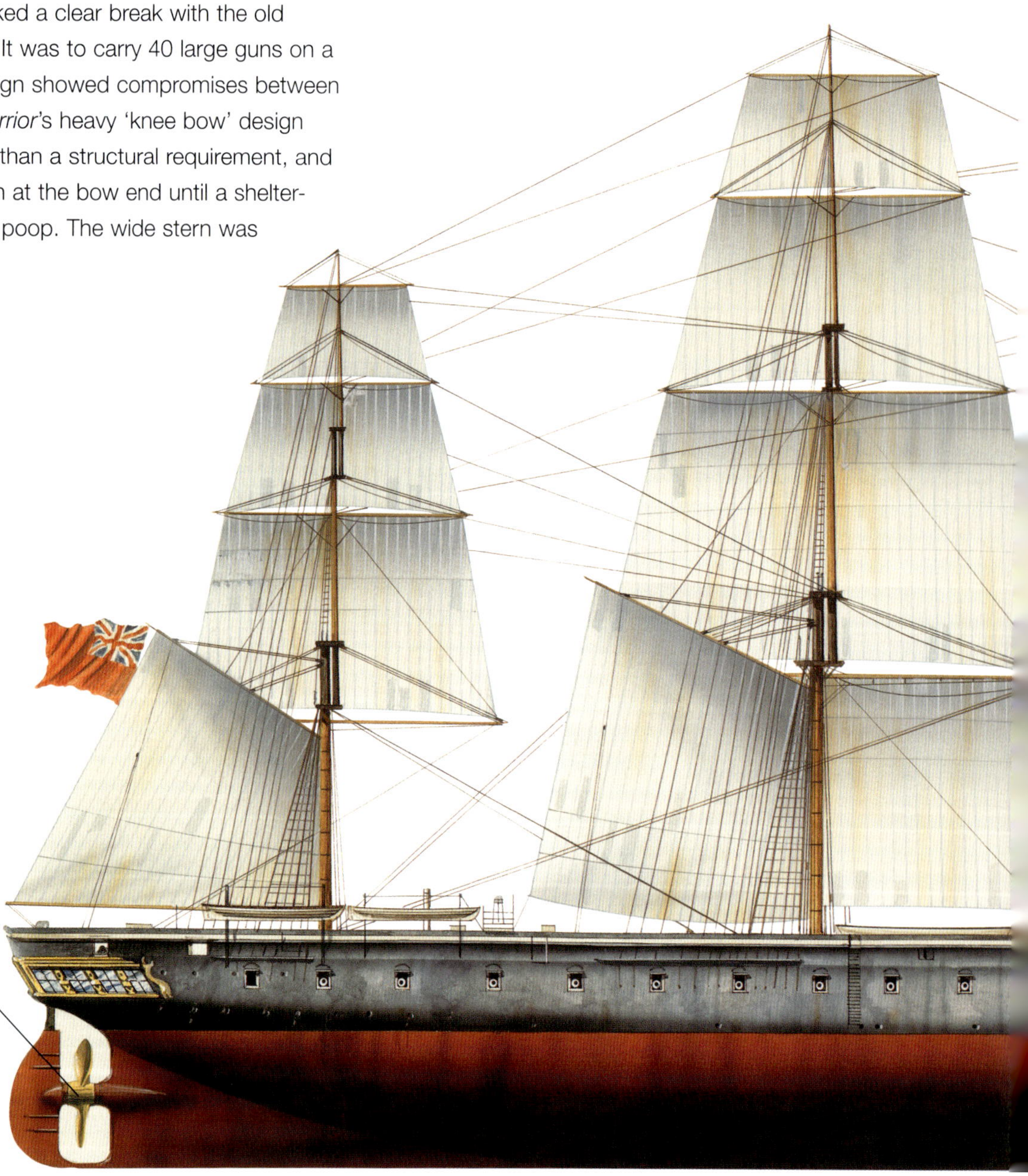

Propeller
Like all screw-driven British warships of the period, *Warrior* had a propeller that could be hoisted up out of the water when the ship was under sail. This required a double stern-post, with the propeller mounted between in a banjo frame and joined to the shaft by a detachable coupling.

Funnels

Warrior's two funnels were telescopic, and were normally taken down when the ship was cruising under sail. In October 1861 their height was raised by 1.83m (6ft) to improve the draught on the furnaces, pushing up speed slightly (0.3 knot).

Smoke emissions

One of the objections to steam power was that a ship's presence was revealed by its smoke. Smoke also got in the way of signalling. Efforts to reduce smoke emission had little effect. High-grade 'steam coal', though expensive, was the best answer.

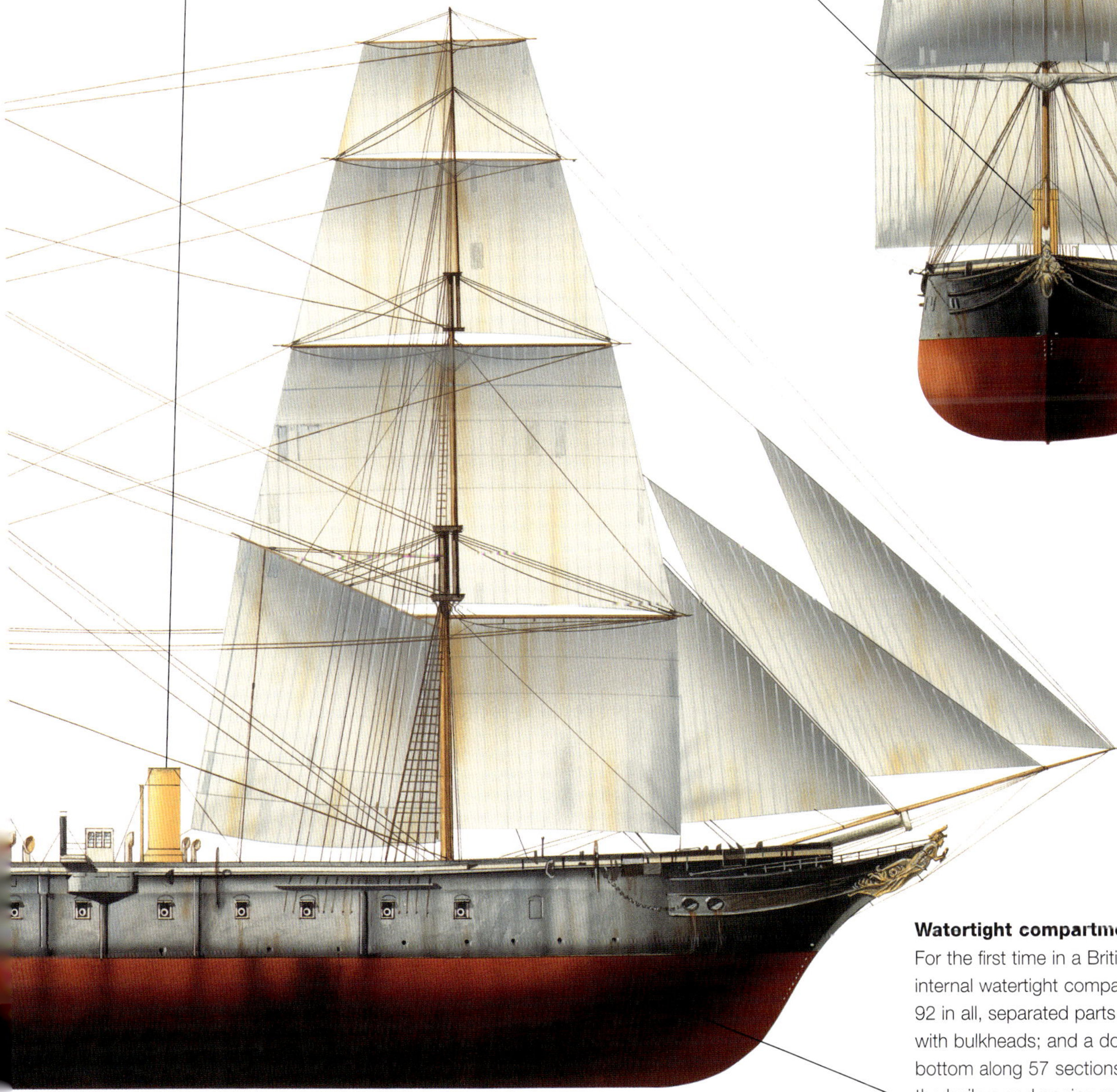

Watertight compartments

For the first time in a British warship, internal watertight compartments, 92 in all, separated parts of the hull, with bulkheads; and a double bottom along 57 sections under the boilers and engine room.

simply copied from existing sailing frigate models. *Warrior* was one of the last three Royal Navy ships to be given a carved figurehead. Another traditional feature, not necessary on an ironclad, was the solid timber bulwarks surrounding the upper deck.

The teak-backed armour plating was applied to the midships section only, 64.9m (213ft) long and 8.23m (27ft) vertically, with 1.83m (6ft) below the waterline. Two bilge keels were fitted to reduce any propensity to roll. The Penn horizontal trunk engine was the most powerful yet installed on a warship.

Third-rate

The new ship presented a problem of classification: its single gun deck defined it as a frigate but as it was expressly designed to overtake and defeat any existing warship, that was clearly inappropriate. The solution was to use its 707-man complement (equivalent to that of a third-rate) as a reason to classify it and its sister ship *Black Prince* as third-rates. The 114mm (4.5in) steel plating was impenetrable by any naval gun of the time (but by 1863 guns had been introduced to pierce such armour). *Warrior*'s 103.6m (340ft) length was notable: great length was identified with the ability to go fast. Sir Baldwin Walker, Controller of the Navy, considered that speed was of the utmost importance 'and absolutely essential in seagoing ships cased with iron'. Care had been taken in designing the underwater lines – even under canvas alone, *Warrior* recorded 13 knots under plain sail and stunsails, and under combined power on 15 November 1861 it made 16.3 knots.

Auxiliary equipment

Warrior was still a broadside ship, its guns arranged in traditional fashion facing outwards. In 1867 the ship was completely re-gunned, with 24 178mm (7in) MLR, 4 203mm (8in) MLR, and 4 9kg (20lb) breech loaders. The Admiralty

The Armstrong breech-loading 110 pounder cannon (seen here) were replaced by four 203mm (8in) muzzle-loaders in 1867.

Black snake

Warrior is a ship of imposing appearance. The hull was painted black, unlike the chequered style of sailing frigates, and seemed immensely long to contemporaries. The great space between foremast and mainmast is very noticeable. When it and the similarly painted *Black Prince* joined the Channel Fleet they were described as 'two black snakes among the rabbits'. They were the only British ironclads with wooden lower masts and caps, though at an early design stage four or even five iron masts had been considered. The original bowsprit was 14.9m (49ft) long but because of the ship's excessive weight in the forward section it was halved in length in March 1862, and the full head-gear was not restored until the poop deck was added between 1872 and 1875, giving the vessel a better balance. Total sail area was 4496.5m^2 (48,400sq ft) including stunsails.

was slow to take advantage of steam power for auxiliary equipment. Originally the only extra was a steam pump. The two-bladed, 9 tonne (10 tons) screw (the largest hoisting screw ever in service) was said to need 600 men to hoist it using the sheerlegs mounted above the double sternposts. *Black Prince* received a steam capstan before *Warrior*, geared to the pump engine. Prior to that, 90 men were needed to work the main capstan. Steering gear was done with a fourfold handwheel set abaft the mizzen mast, and directed from a low bridge mounted on the quarter-deck bulwarks, with an armour-plated blockhouse structure beneath. But the steamships' length had an adverse effect on manoeuvrability. The Navy Controller's specification required only three turns of the wheel to give a full degree of helm, as with narrow-ruddered sailing ships; this made steering of a long screw-driven ship an immensely heavy task (a fourth turn was conceded in 1861 for steamships of more than 298kW (400hp). *Warrior* carried two bower and two sheet Admiralty-pattern wooden-stocked anchors, each weighing 4.3 tonnes (4.75 tons), 1.42 tonne (1.67 tons) iron-stocked stream anchor, and two iron-stocked 0.87 tonnes (0.96 tons) kedge anchors.

Specification

Dimensions	Length 127m (418ft) Displacement 8355 tonnes (9210 tons)
Propulsion	Penn horizontal single trunk of 3930kW (5270hp); 3 masts, ship-rigged
Armament	26 68-pounder, 10 110-pounder, 4 70-pounder guns
Speed	14.08 knots
Complement	707

Years of service

In 1904 it was a torpedo school and later an oil hulk before restoration to original condition in the 1980s. It is now a museum ship at Portsmouth. With *Black Prince* it towed a floating dock across the Atlantic from Madeira to Bermuda in 1869. Another refit between 1872 and 1875 saw it provided with the poop deck and steam capstan, and it served as coastguard ship at Portland, then from 1881 to 1884 as a training ship for reservists on the Clyde. In 1881 it was re-classified as an armoured cruiser. In 1904 it was adapted for service with the HMS *Vernon* torpedo school and the cut-down hull was finally transferred to Pembroke to provide a pier for an oil pipe-line.

König Wilhelm (1869)

At first the flagship of the Prussian fleet, after 1871 it became the largest battleship in the navy of the newly-unified German Empire, and retained that status for 20 years, until Germany's naval expansion got under way.

Classified by the Prussian Navy as a *Panzerfregatte* or armoured frigate, like other such vessels of the time, *König Wilhelm* was in effect a battleship, the 'frigate' being a now anachronistic reference to the single gun-deck, which by now carried more firepower than any former three-decker. It had been ordered by the Imperial Ottoman Navy from the Thames Ironworks at Blackwall, London, but when the Turks cancelled the order, the Prussians stepped in and bought it on the stocks in February 1867.

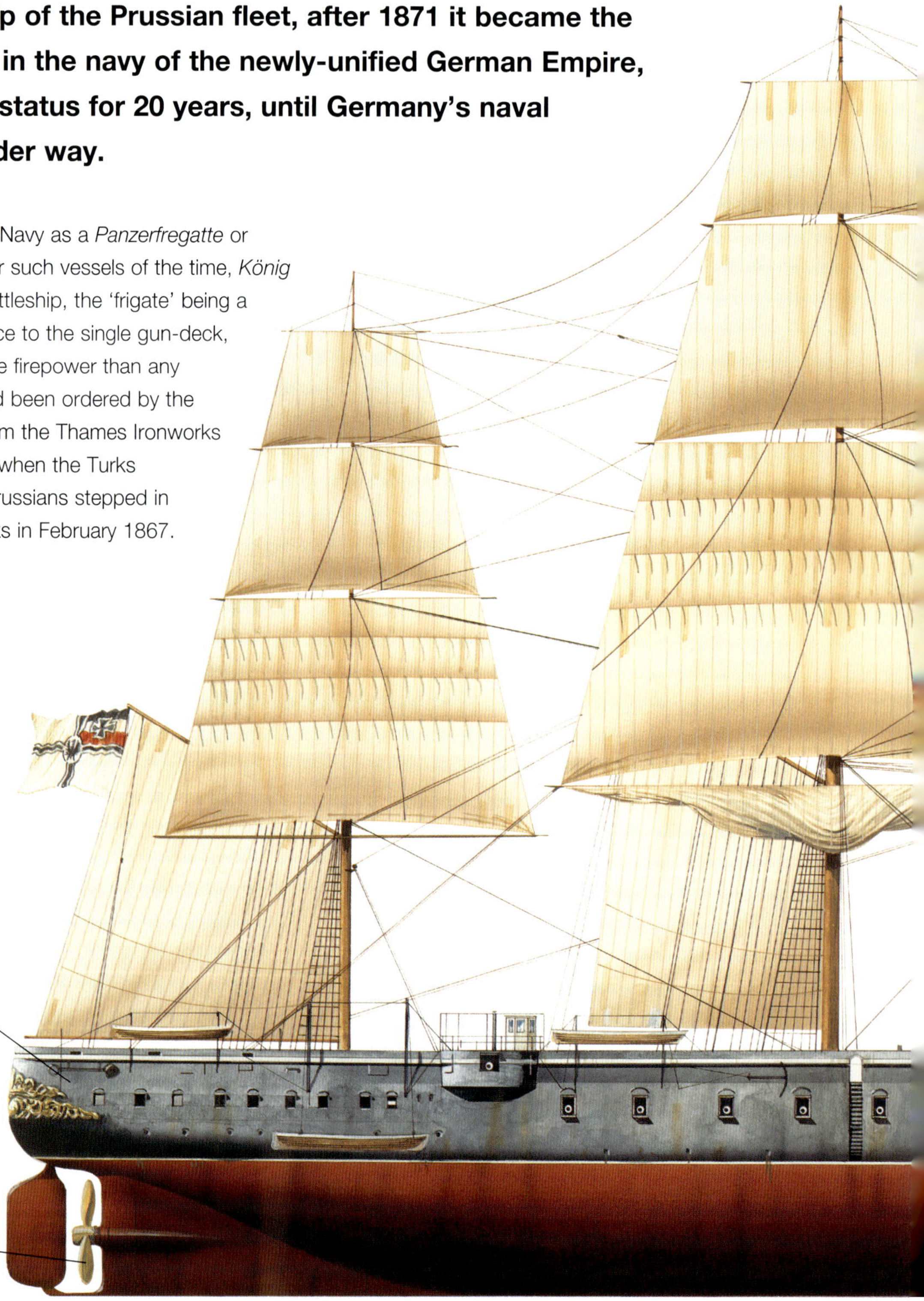

Plates
Wrought-iron plates were fixed to a teak backing. At the midships waterline, plate thickness was 305mm (12in) on the outer armour, with 178mm (7in) hull plates and 250mm (9.8in) of teak.

Single screw
The single screw was four-bladed, of 7m (23ft) diameter.

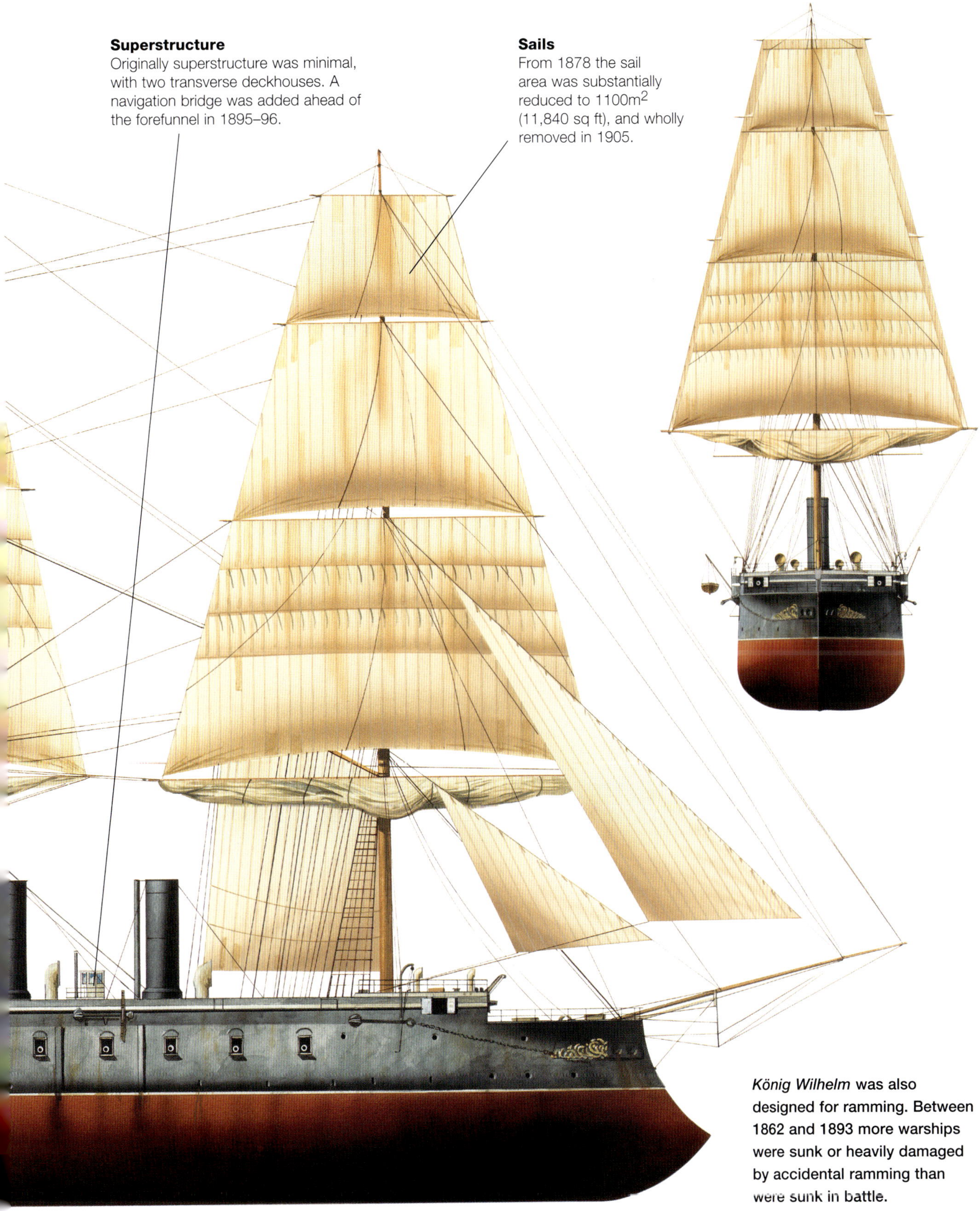

König Wilhelm was also designed for ramming. Between 1862 and 1893 more warships were sunk or heavily damaged by accidental ramming than were sunk in battle.

König Wilhelm

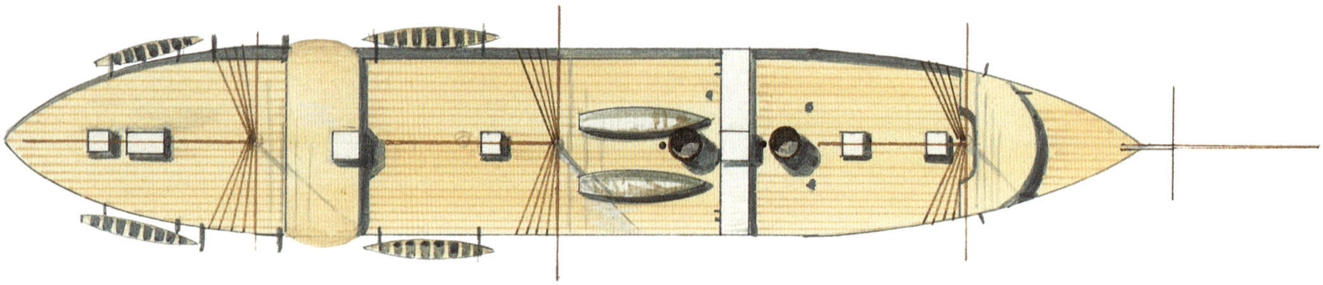

Launched on 25 April 1868, the ship was named after King Wilhelm I.

A deck plan of the ship in its original form. Despite the high freeboard, the weather deck is almost entirely clear of equipment. The battery is on the main deck, below.

Specifications and armament

The hull was fitted with the ram-type bow fashionable at the time, though a lengthy bowsprit extended before it, housed in a 'cubby' at a lower level than the weather deck. It was a high-sided ship, with 12.94m (42ft 5in) of freeboard. The hull was divided into 11 watertight compartments, and was double-bottomed for 70 per cent of its length. Steam power came from a Maudslay horizontal two-cylinder engine. Eight boilers were set in two rooms, each with a funnel, and the three masts could mount a spread of 2600m^2 (27,986 sq ft) of sail.

The original armament, which would quickly become obsolete, was 33 muzzle-loading 32-pounders, smooth-bored. The choice of guns was presumably based on British advice. At this time the Admiralty's policy on guns was in some disarray. Armstrong breech-loading guns with rifled barrels, intended for armour-piercing, and introduced in 1860, had been a failure, especially the 110-pounders. Discussion of the respective merits of smooth-bore and rifled bore, of loading by breech or muzzle, was intense, and various designs were tried.

In 1865 a 64-pounder muzzle loader was introduced using 'Woolwich' rifling, devised at the Royal Gun Factory, but it was not a success. It would be 1880 before the British finally resolved their naval gun problem.

Grand review

It became customary in the age of the steam-powered capital ship to use such vessels for peaceful visits to countries allied or at least not hostile. These courtesy visits were also opportunities to show off the latest in the way of naval design. Naval reviews, with long lines of ships assembled, and visiting warships from other nations, were a favourite way of demonstrating British imperial power and the might of the Royal Navy.

To the grand review organised at Spithead (Portsmouth) in 1896 to celebrate Queen Victoria's diamond jubilee, Kaiser Wilhelm II, her grandson, sent *König Wilhelm*, rather than the new *Brandenburg*, to represent the Imperial German Navy.

It was perhaps a double-edged compliment to the host-nation, to send a ship originally built in Britain, but radically modernised and improved in Germany: an intimation that Germany had every intention of rivalling Great Britain as a first-class naval power.

König Wilhelm in its original pre-1896 form. It appears that the funnels were retractable, as with some other warships of the period, when the engines were not in use.

Imperial fleet

Prussia was a kingdom with strong military ambitions. In 1870 France declared war on Prussia only to be quickly defeated on land. *König Wilhelm* was flagship of the Prussian fleet but (despite the much greater size of the French Navy) naval forces played little part in the war, and *König Wilhelm* was immobilised by engine problems for much of the time.

In 1871 the German Empire was proclaimed, King Wilhelm became Kaiser, and it became the largest ship of the Imperial Fleet, which was simply the old Prussian one, though a new building programme was put in hand. On 31 May 1878 *König Wilhelm* accidentally rammed and sank one of the new vessels, the turret ironclad *Grosser Kurfürst* (commissioned in 1875) in a faulty manoeuvre to avoid a couple of sailing craft as they cruised up the English Channel off Folkestone. As one British admiral commented, 'the ram was more dangerous in peace than in war'.

Modernisation

A complete modernisation was done in Hamburg in 1896. The sail rig was taken down and replaced by two new-type naval masts, with observation platforms, fighting tops with light guns, and signalling arms. A small third mast remained, with a bridge house built in front of it. The cubby where the bowsprit had been stepped was plated over, though the original decorative scrollwork was left in place, and a forward lookout and searchlight platform was installed.

New boilers and triple-expansion engines were installed, and the ship was completely re-gunned. Twenty-two Krupp 239mm (9.4in) and one 152mm (6in) rifled breech-loading guns were installed and supplemented by 18 88mm (3.5in) quick-firing guns. Five torpedo tubes were fitted, above the waterline, to fire weapons scarcely anticipated when it was originally planned.

By 1893–94 the German Navy's four new *Brandenburg*-class battleships were being commissioned, and in its new form *König Wilhelm* was redesignated as a *Grosser Kreuzer* (armoured cruiser). From 1893 to 1897 it was flagship of Division II of the Imperial German Navy, at the new North Sea base of Wilhelmshaven. After 1897, as the German fleet grew in size and power, it was relegated to a harbour defence and training role until its removal from the active list in 1904. Disarmed, it continued in use as an accommodation ship, at Kiel, and did not go finally out of service until 1921, when it was scrapped at Rönnebeck, near Bremen, having outlasted the German Empire by three years.

Specification

Dimensions	Length 112m (367ft 5in), Beam 18.3m (60ft), Draught 8.56m (28ft 1in), Displacement 8711 tonnes (9603 tons)
Propulsion	(1869): 8 boilers, Maudslay horizontal 2-cylinder engine developing 6000kW (8000hp)
Speed	14.5 knots
Armament	(1869): 33 72-pounder ML guns (smooth bore)
Armour	Belt 305–152mm (12–6in), Battery 152mm (6in), Deck 50mm (2in)
Range	(after 1896): 4148km (2240nm) at 10 knots
Complement	730

Devastation (1873)

HMS *Devastation,* the first seagoing turret ship, carrying the heaviest guns yet, was a highly controversial design. But its big-gun layout, and its reliance on engines only, without sails, proved successful against the critics' prophecies of doom, and gave many pointers to further development.

Devastation was the creation of Sir Edward Reed, Chief Constructor at the Admiralty since 1862, and closely involved in the technical, but also the financial and political, aspects of warship design in a period of radical change and development. Numerous post-1870 alterations were made to the design by his successor Sir Nathaniel Barnaby.

Freeboard
The freeboard of the low armoured deck was no more than 1.37m (4ft 6in) above the waterline, but higher than American monitors and calculated to give an effective margin of buoyancy.

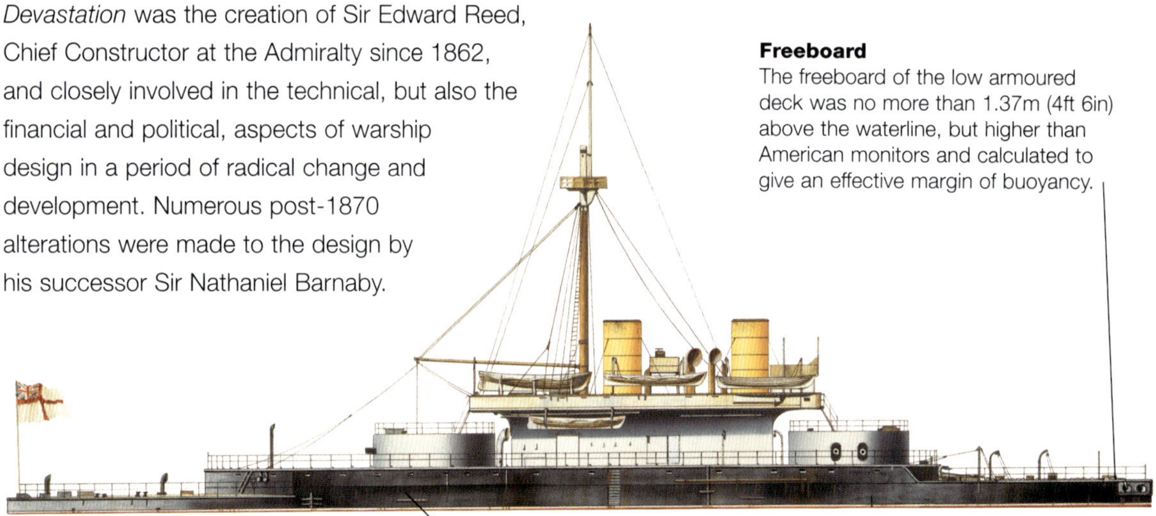

Hull sides
The hull sides were parallel for two thirds of the ship's length, another original feature in an age when warships' sides normally curved from stem to stern.

ML guns
The ML guns were turned round completely and angled downwards for loading: a procedure which slowed down their rate of fire to one round per two minutes.

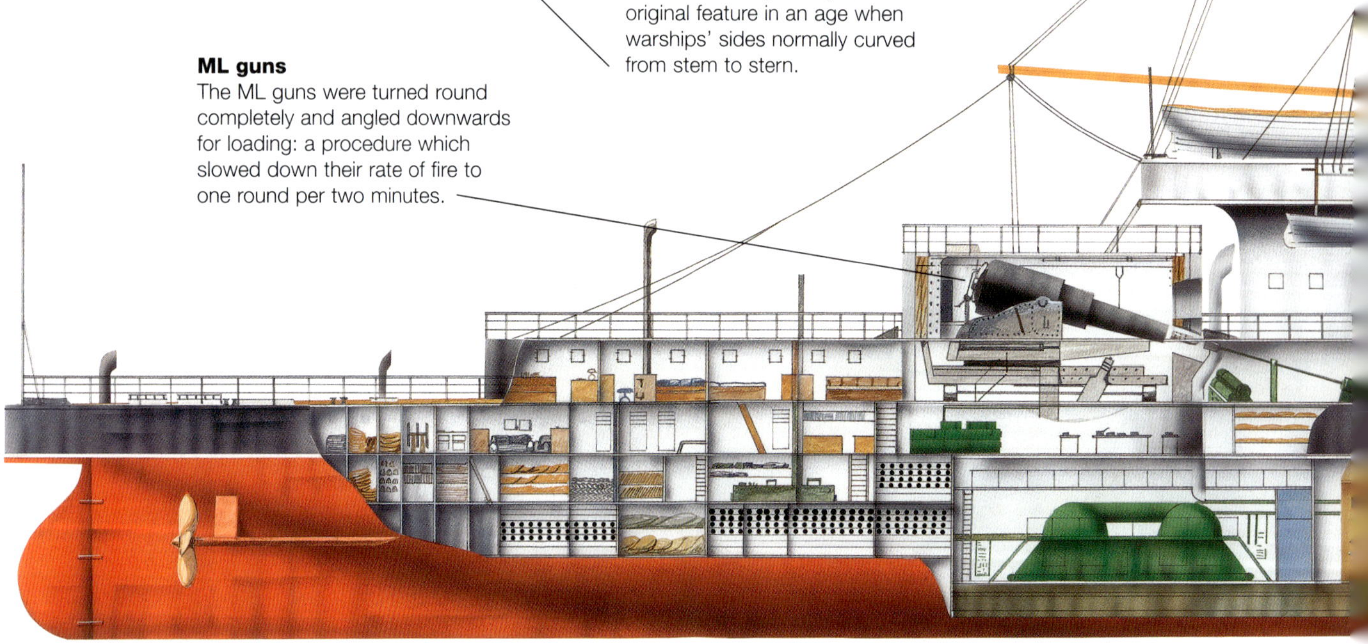

The turret ship had its ancestry in the American ironclad *Monitor* of 1862. Under Reed a British 'breastwork monitor', HMS *Cerberus,* was built as a guardship for the port at Melbourne, Australia, in 1870; and *Devastation* was in most ways an enlarged, seagoing version.

The low freeboard of a monitor-type ship improved steadiness and required less armour than a higher-sided ship. The armoured breastwork provided a degree of protection to the deckhouse and the bases of the two twin-gun turrets, mounted in-line on the centreline each with angle of fire of 180 degrees. These turrets, unlike a broadside-mounted battery, gave the ship 360 degrees of firepower. With heavy guns, such a ship was far more flexible in battle. The monitor-form also had disadvantages, including the fact that it might be seaworthy but often had to proceed in a manner more like a submarine, with decks awash and all hatches shut tight against the water. A stormy transatlantic voyage to Britain by the American double-turreted *Miantonomoh* in 1866 was a desperate ordeal for its crew.

To observers in the 1870s, the appearance of *Devastation* was extraordinary, with its low freeboard, substantial superstructure, and the two massive turrets.

Naval sensation

Devastation was laid down at Portsmouth on 12 November 1869, launched on 12 July 1871 and completed on 19 April 1873, at a total cost of £361,438. The capsizing of

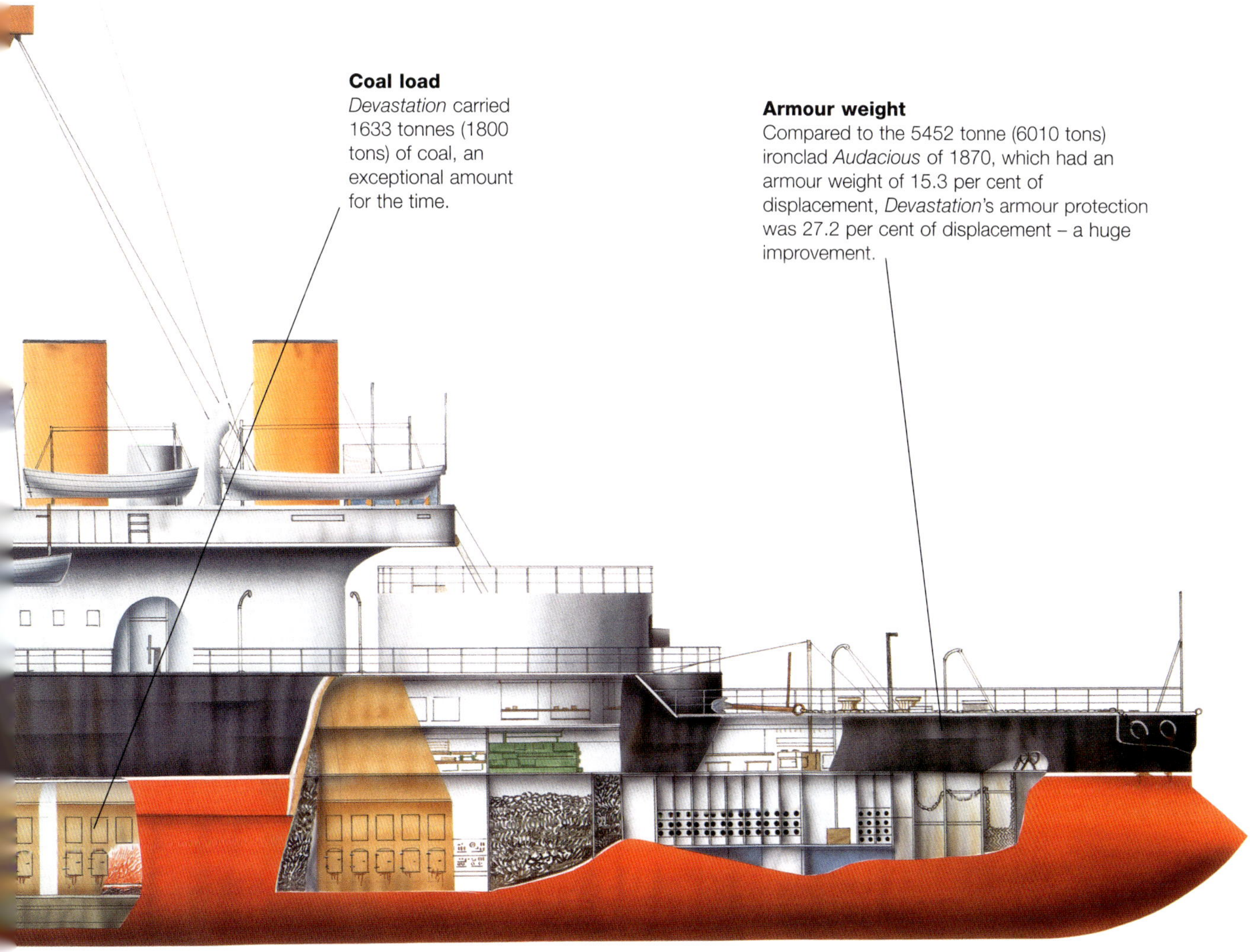

Specification

Dimensions	Length 86.9m (285ft), Beam 19m (62ft 4in), Draught 8.4m (27ft 6in), Displacement 8464 tonnes (9330 tons)
Propulsion	Penn direct acting trunk engines of 4959kW (6650hp), 2 screws
Armament	(1873): 4 31.7-tonne (35 tons) 305mm (12in) MLR guns
Armour	Belt 305–216mm (12–8.5in), Breastwork 305–250mm (12–10in), Turrets 356–250mm (14–10in), Conning tower 228mm (9in), Deck 76–50mm (3-2in)
Range	8704km (4700nm) at 10 knots
Speed	13.84 knots
Complement	358

the turret ship HMS *Captain* with the loss of almost all its crew in September 1870 held up construction, though *Captain*'s instability had been caused by inflicting three masts and a full set of sails on the design.

On its commissioning, it was as much of a sensation as HMS *Dreadnought* would be 30 years later, looking utterly unlike any previous British man of war. The freeboard was only 1.37m (4ft 6in), allowing waves to wash over the weather deck. The hull was virtually square in section, a vast box with rounded ends, and a short ram was placed below the waterline.

Internally the hull was divided into 68 compartments. It had a double bottom, with the space between capable of holding 907 tonnes (1000 tons) of water ballast. A central superstructure supported the two funnels, with a low, armoured pilot tower placed between them. At each end, protected by an armoured breastwork, was a massive gun turret, 9.3m (30ft 6in) in diameter, with two 305mm (12in) ML guns firing shells of 320kg (706lb).

Most striking perhaps, there was a single mast and no sailing rig at all. Two boiler rooms and the engine occupied the central part of the hull. Penn's direct-acting trunk-type engines were fitted, for the last time in a British battleship, driving twin four-bladed propellers. Despite forecasts of

Comparison with some of the previous illustrations will show how very unlike *Devastation*'s appearance was to the established notion of how a battleship should look in 1873.

Armament

This was the first complete application to a seagoing battleship of the principle of mounting the guns on top of the hull rather than inside it. *Devastation*'s original 305mm (12in) guns were the largest muzzle-loaders ever mounted in a ship. The turrets each had an arc of fire of 280 degrees. But their range was limited and the rate of fire was poor. In 1886 the Royal Navy readopted the principle of breech loading, and the 254mm (10in) breech-loading replacement guns of 1892, made by Whitworths, firing 227kg (500lb) shells, had double the range and three times the rate of fire. The Nordenfelt guns, added in 1879, were multiple-barrelled machine guns patented in 1873, operated by one man, and produced in various calibres from rifle-size to 25mm (1in). The Royal Navy used them as anti-torpedo boat weapons. The Nordenfelt company was taken over by the Maxim Gun Co. in 1888.

disaster, *Devastation* behaved in rough seas exactly as its designers had predicted, and proved to be a steadier gun platform than the long-hulled sail-rigged vessels.

Refit

In 1879 *Devastation* underwent a major refit, with six Nordenfelt guns mounted on the hurricane deck and a Gatling gun placed in a fighting top on the mast. By now the torpedo was part of the naval armoury and ports for 14in (356mm) torpedoes were cut in the breastwork armour on each side. Twelve torpedoes of Whitehead design were carried. Further modifications in 1891–92 included rearming with 254mm (10in) breech-loading guns, and replacement of the Nordenfelt guns with six 6-pounders and eight 3-pounders. New inverted triple expansion engines were installed with cylindrical tubular boilers, and funnels of up-to-date type. An enclosed flying bridge was constructed to improve viewing.

'Fighting coal mine'

Though *Devastation* was never engaged in a battle, its performance in exercises confirmed the merits of the design. With its large bunker capacity (reduced in 1892 with new and more efficient machinery) it was described as a 'fighting coal mine'. Attached to the Channel Fleet in 1874–75, it transferred to the Mediterranean until November 1878. After refit it remained in reserve until April 1885, and from August 1885 to May 1890 was port-guard ship at Queensferry. After completion of the second refit (1892) it had spells as port-guard ships, including Gibraltar from November 1898 to April 1902 and on reserve. Stricken in 1907, it was sold for scrapping in 1908.

Italia (1880)

A novel battleship design, with four very large guns and no side armour, in many ways *Italia* was a forerunner of the battlecruiser. For a few years it held the prestige position of the largest and fastest battleship in the world.

The Kingdom of Italy was declared in 1861 and from the start it had difficult relations with the French and the Austrians. For Italy with its long coasts and numerous islands, a naval force was a prime necessity, and by 1866 it had fought one of the first 'modern' sea battles, against the Austrian Navy at Lissa in the Adriatic Sea. That was a defeat despite superior numbers on the Italian side, and drove the Italians to further expansion of their fleet.

Fast and powerful

Laid down at Castellamare in 1877 and launched in 1880,

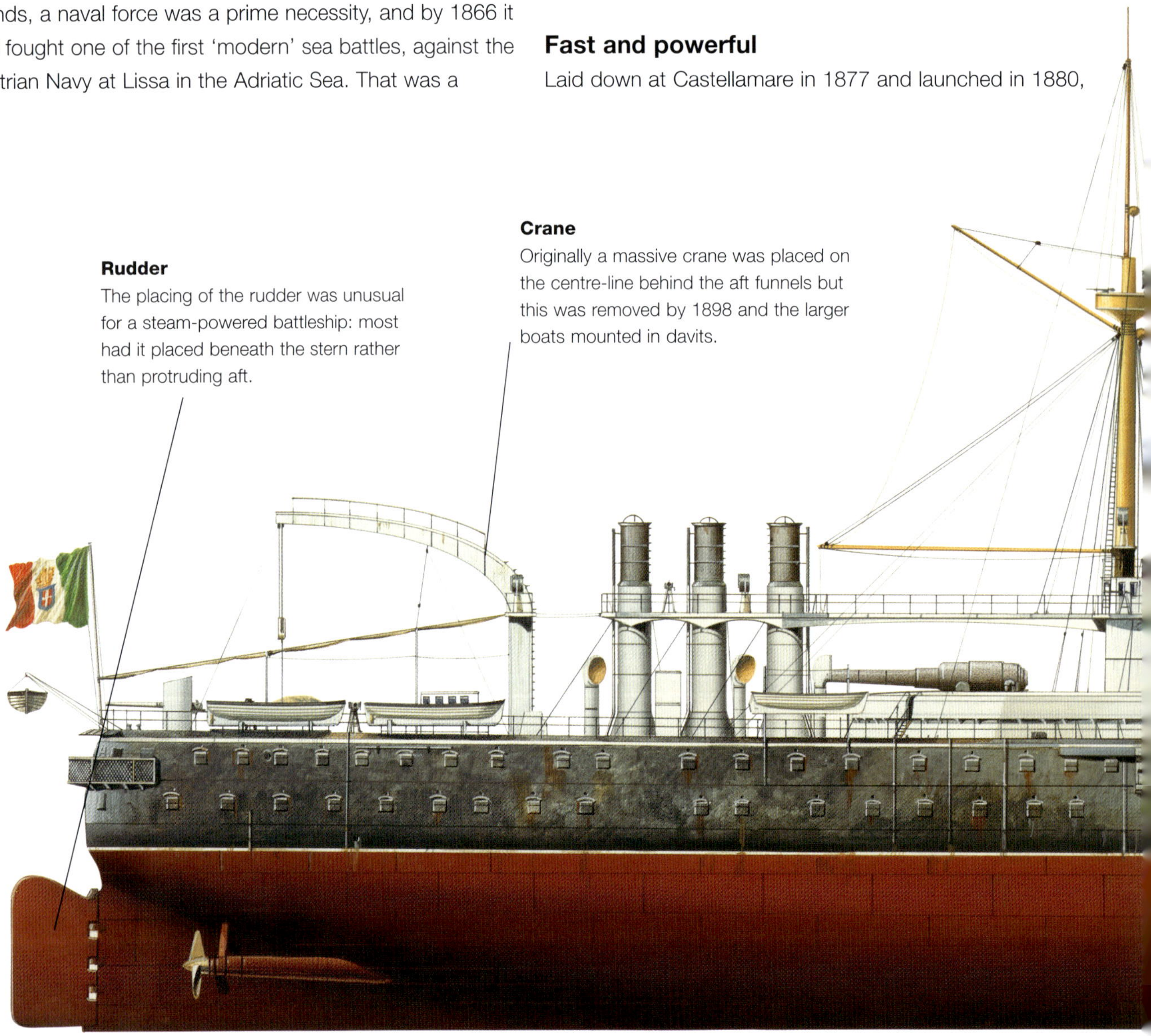

Crane
Originally a massive crane was placed on the centre-line behind the aft funnels but this was removed by 1898 and the larger boats mounted in davits.

Rudder
The placing of the rudder was unusual for a steam-powered battleship: most had it placed beneath the stern rather than protruding aft.

Italia took Brin's revolutionary design of *Duilio* (1876) to an extreme. The brief was for a very fast ship, heavily-armed, which could also carry a large number of troops (at the time, France and Italy were on the verge of war over Tunis, on the south coast of the Mediterranean Sea). As with *Duilio*, the guns were mounted in echelon, but the main armament of *Italia* and its sister ship *Lepanto* was of even heavier calibre, four 432mm (17in) guns each weighing 93 tonnes (103 tons), firing shells of 907kg (2000lb). The guns were mounted in a huge barbette of oval shape extending beyond the sides, forming an armoured redoubt set

The frontal view shows the flying bridge, the elevated forward control and navigation house, the sponsons, and the rounded hull shape.

Guns
Unlike HMS *Inflexible, Italia's* big guns were not enclosed in turrets but rose above the rim of the barbette.

Light guns
In 1898 11 lighter quick-firing guns were added, on the forecastle and quarter-decks, as anti-torpedo boat defences.

diagonally across the hull. Unlike the British *Inflexible*, it had a high freeboard, 7.6m (25ft), offering more of a target to an enemy. The sides carried no armour, but *Italia* relied on the power of its guns, and its high speed, to avoid attack. Six funnels in sets of three, linked by high catwalks with the conning tower, a lofty central mast, and a large curved crane on the afterdeck, gave *Italia* a unique appearance. One of the few more traditional features was an 'admiral's walk' around the curved stern.

Fittings

The ship was built largely of steel, rather than iron. Internally it had the now-standard armoured deck, curving upwards slightly from the sides 1.83m (6ft) below the waterline, but above it a cellular raft ran the entire length of the ship. The space between them was lined laterally with cork-filled watertight cells separating the hull plating from an inner cofferdam on each side, and two transverse levels, one of empty cells, with coal storage space below. A double bottom was also fitted.

One novel feature of *Duilio*, not maintained on the new ship, was a stern compartment for a torpedo boat, secured by watertight doors. *Italia* also had space to hold an infantry division of 10,000 men and its equipment for the relatively short Mediterranean crossing. The main guns could be independently trained and aimed, but as with other very large guns of the time, the rate of fire was slow, no more than one round every four or five minutes.

Construction of *Italia* and *Lepanto* stretched Italy's new warship-building resources, and the Italian government did not proceed to enlarge its battlefleet further. But the size, speed and general innovation of these Italian capital ships had a major impact on ship design and naval planning in both the British and the French navies. Sir Nathaniel Barnaby, Britain's chief naval designer, observed that 'We must ... regard the first-class ironclad as ... being of over 14,000 tons if we accept the reasonings of the Italian architects and the expression of their ideas in the *Italia* and *Lepanto*'. In the mid-1880s British designers were still

Specification

Dimensions	Length 124.7m (409ft), Beam 22.5m (74ft), Draught 8.7m (28ft 8in), Full load 10.1m (33ft)
Displacement	15,900 tonnes (15,654 tons)
Propulsion	24 boilers, 2 vertical compound engines developing 11,780kW (15,797hp), twin screws
Armament	4 432mm (17in) breech-loading guns of 93 tonnes (103 tons), 7 150mm (5.9in) and 4 119mm (4.7in) guns, 4 356mm (14in) torpedo tubes
Armour	Redoubt 483mm (19in), Boiler uptakes 406mm (16in), Conning tower 102mm (4in), Deck 102–76mm (4–3in)
Range	9260km (5000nm) at 10 knots
Speed	17.8 knots
Complement	701

From this deck plan the redoubt's diagonal arrangement is clear. There was an armoured deck just below waterline level, but no citadel or side armour.

mulling over the kind of ship 'most suitable for meeting the *Italia*'. In this way the Italian contribution was to push the greater naval powers towards greater size.

Rebuild

Between 1905 and 1908 *Italia* was rebuilt, losing two funnels and with the tall single mast replaced by two, forward and aft of the funnels. By this time battleship development had caught up and moved on. Improved armour had disproved Brin's theory that gun-power had made side-armour pointless, and the formidable guns were sadly out of date. By the 1890s the ship really ranked with armoured cruisers. The secondary armament was changed and reduced in quantity. In 1909–10 it was used for torpedo training. Still in commission during World War I, but renamed as *Stella d'Italia*, it was based at Taranto and Brindisi for gunnery training until 1917, when it was disarmed and transferred to the mercantile marine as a grain transport. It was returned to the Regia Marina in 1921, but was almost immediately sold for scrapping.

Battle of Lissa

The Battle of Lissa, the first major battle between ironclad warships, was fought on 20 July 1866, off the island of Lissa (Vis) in the Adriatic Sea. Although the Italian fleet was numerically stronger, with twelve ironclads compared to the Austrians' seven, its organisation was poor and the Austrian Admiral Tegetthoff, on the flagship *Erzherzog Ferdinand Max* displayed better tactics than the Italian Admiral Persano on *Affondatore*, isolating the Italian ships in groups, so that in the main engagement, he could pitch his seven ironclads against four Italians. Lissa was fought at a period when armour development had temporarily outpaced advances in gunnery.

Much of the fighting was conducted at close quarters, and both sides engaged in ramming opposing ships. This aspect of the battle was against the long-term technical trend towards longer-range gun-based conflicts, but resulted in most heavy ships being given ram-type bows for the next 40 years.

Inflexible (1881)

HMS *Inflexible* was, like HMS *Warrior* in 1859, intended to emulate and surpass a new foreign battleship design; in this case the Italian ships *Duilio* and *Dandolo,* the most heavily-armoured and fastest warships of their time, with massive 90-tonne (100 tons) guns in a central citadel.

For Great Britain to lose its naval supremacy was unthinkable, and the Italian ships, designed by Benedetto Brin, sent British warship designers back to their drawing boards to produce a ship of outstanding power, speed and size. It took some time. Laid down at Portsmouth on 24 February 1874, *Inflexible* was launched on 27 April 1876

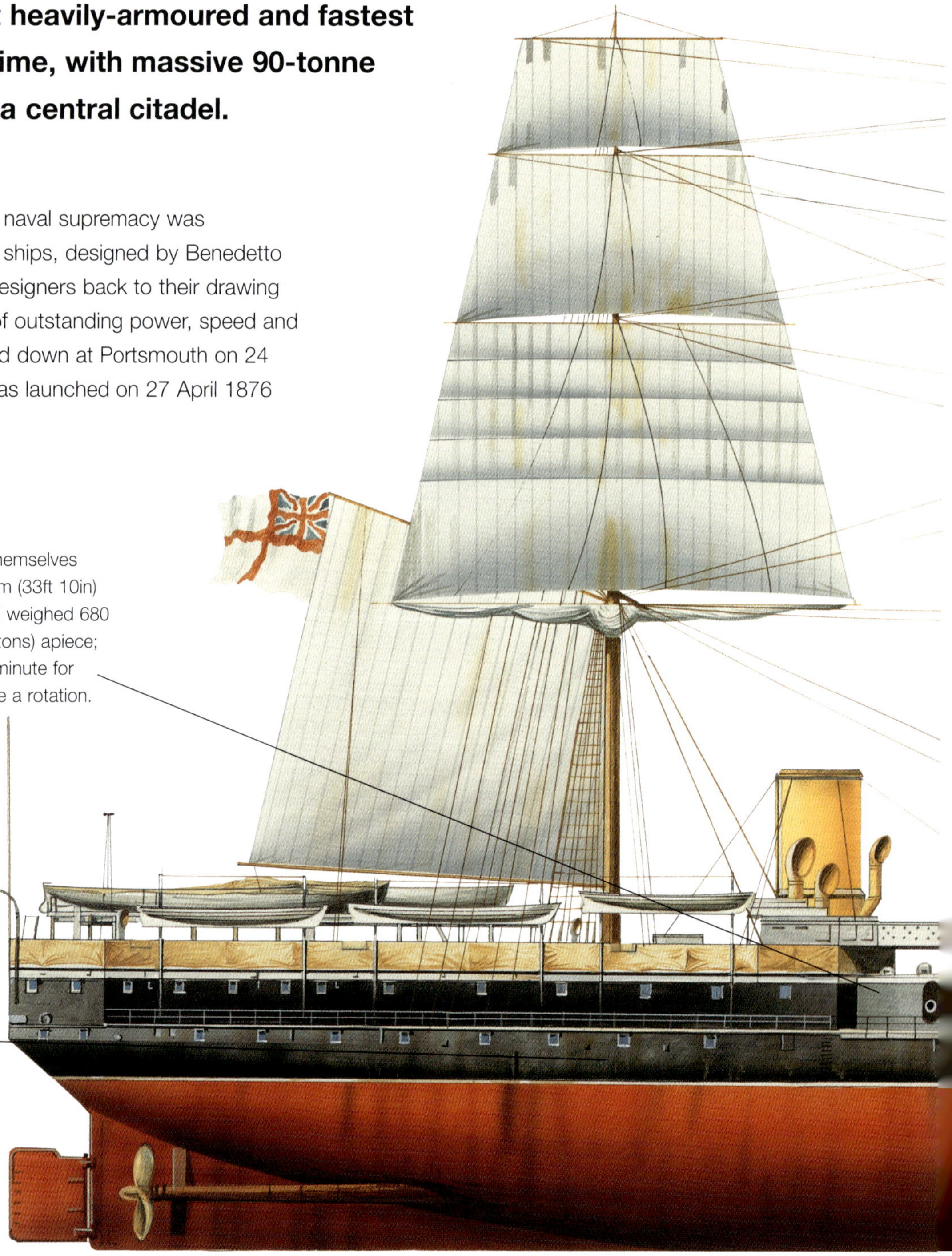

Turrets
The turrets themselves were of 10.3m (33ft 10in) diameter and weighed 680 tonnes (750 tons) apiece; it took a full minute for them to make a rotation.

Armour plating
The armour anticipated the later American 'all or nothing' system, with very heavy protection round the central citadel (a third of the hull length) protecting magazines and machinery, while the ends of the ship were unarmoured.

Inflexible was the only turret ship of its time to carry full rigging. The sail area was 1719m² (18,500sq ft). By comparison the tea clipper *Cutty Sark* carried a sail area of 2976m² (32,000sq ft).

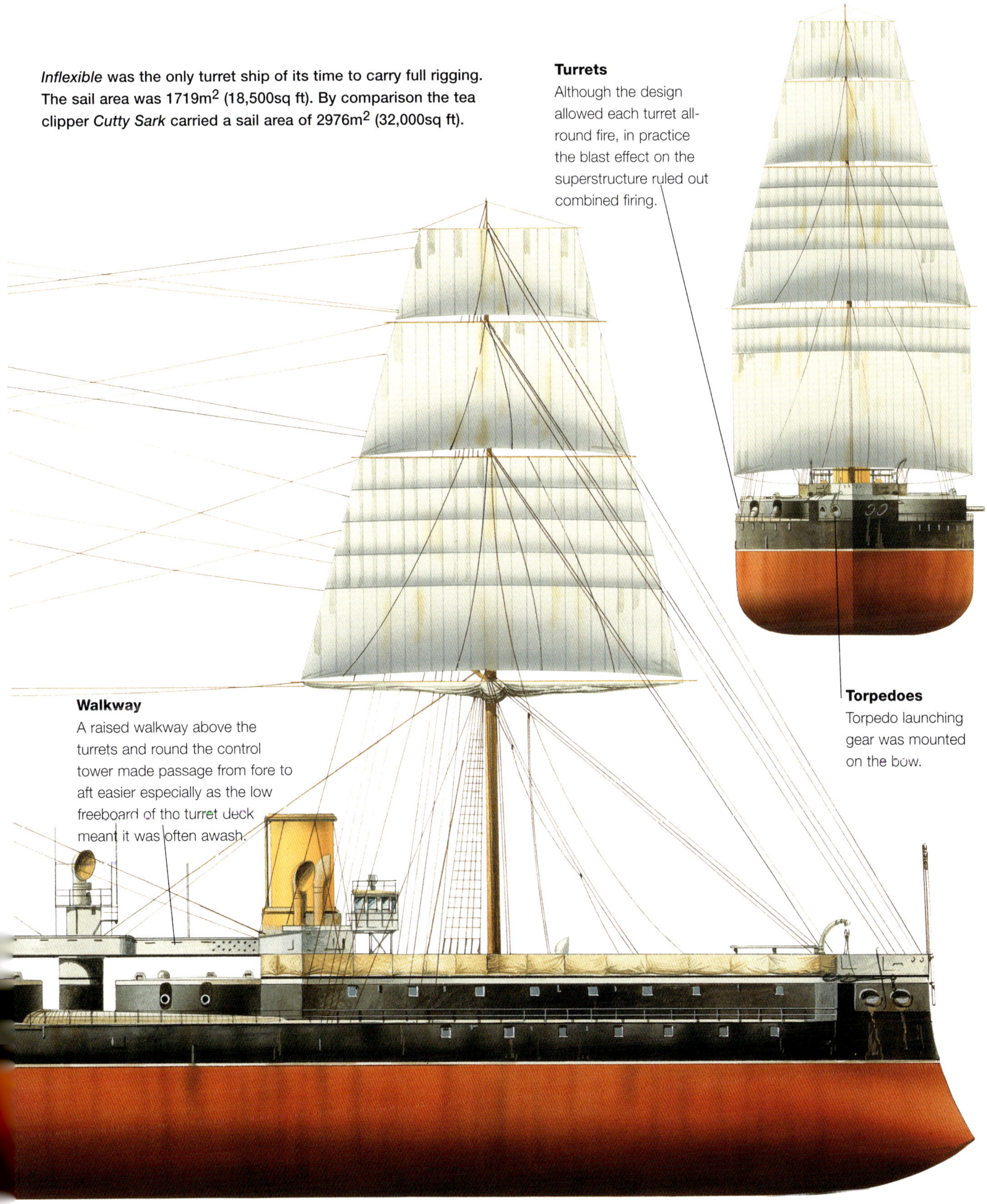

Turrets
Although the design allowed each turret all-round fire, in practice the blast effect on the superstructure ruled out combined firing.

Torpedoes
Torpedo launching gear was mounted on the bow.

Walkway
A raised walkway above the turrets and round the control tower made passage from fore to aft easier especially as the low freeboard of the turret deck meant it was often awash.

but not completed until 18 October 1881. Its cost was £812,485.

Specification

The gun arrangement, modelled on *Duilio*'s, was of two large twin central turrets mounted in echelon formation, the port one set ahead of the starboard one, at opposite corners of a 'citadel' of unprecedented armoured strength, 609mm (24in) at the maximum.

A 76.2mm (3in) armoured deck ran the length of the ship, beneath the main deck and below waterline level, in place of vertical armour along the waterline. Above this deck the hull was divided into many compartments to localise flooding. The citadel was an armoured box rising to 2.9m (9ft 6in) above the waterline and extending 1.8m (6ft) below, enclosing the magazines and hydraulic gear to work the turrets. Below it were two boiler rooms, with the engines set between. The ship was the broadest in relation to length yet built, in order to ensure stability and give the gun turrets the maximum possible arc of fire. The forward and stern sections were not armoured but subdivided into many watertight compartments with coal bunkers on each side and inner walls thickly lined with cork. These parts could sustain damage without affecting the ship's fighting capacity, while the citadel was regarded as invulnerable, designed to stop even 408mm (16in) shells. The superstructure was kept well in to the centre of the ship, in order to maximise the arc of fire for the guns.

At the stern, two single-tube torpedo boats were carried, launched by a derrick from the aftermast, and this was also the first armoured ship to mount underwater torpedo tubes, with two placed experimentally in the bows, though they cannot have been compatible with the ram. Two 356mm (14in) torpedo carriages were mounted on deck.
Among the many other technical advances incorporated were compound expansion engines, and dynamos to light the ship by electricity (the first warship so fitted). In important ways, notably the full-length armoured lower

Specification

Dimensions	Length 97.5m (320ft), Beam 22.9m (75ft), Draught 8.08m (26ft 6in), Displacement 10,777 tonnes (11,880 tons)
Propulsion	12 boilers, 2 Elder & Co. 3-cylinder compound engines, 6269kW (8407hp), 2 screws. Sail area 1719m^2 (18,500 sq ft)
Armament	4 406mm (16in) MLR 72.5-tonne (80 tons) guns; 6 20-pounders; 2 submerged torpedo tubes, 2 torpedo carriages, 356mm (14in)
Armour	Citadel 610–406mm (24–16in), Bulkheads 559–356mm (22–14in), Turrets 432–406mm (17–16in), Deck 76.2mm (3in)
Speed	14.75 knots
Complement	440

HMS *Inflexible* in 1881, before replacement of the original masts and rigging, showing lifeboat davits extended.

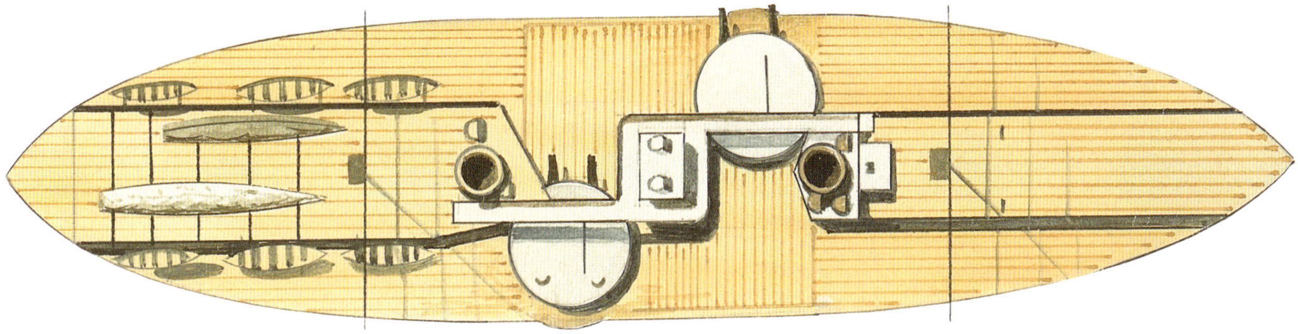

deck, *Inflexible* established standards which later battleships would follow, though the debt to Brin and the Italian ships should not be forgotten. But the citadel formation, though it became popular for a time, was ultimately dropped.

The original plan was to place the funnels almost side by side between the turrets, but this was dropped in case they should be damaged by gunnery blast.

Poor armament

Inflexible's weakest aspect was the guns. The original design had specified guns weighing 54 tonnes (60 tons), but in 1875, with the ship already building, this was changed to 72.5 tonnes (80 tons). Muzzle loaders, of 406mm (16in) calibre, they fired a 764kg (1684lb) shell capable of piercing armour 584mm (23in) thick at a distance of 914m (1000yd). They had to be depressed with the barrel-mouths below a protective armoured cover for the shells to be inserted from a hydraulic ram, then re-aimed. Firing was limited to one round every two minutes. And within a few years the massive muzzle loaders were outclassed by lighter, faster-firing breech-loading heavy guns.

A refit was done in 1885. The two masts had been rigged as for a brig, though this was essentially cosmetic so that men could man the yards for display purposes. In the refit, the rig was abandoned for two pole masts.

Mediterranean Fleet

The appearance of *Inflexible* has been condemned as graceless, tubby and low in the water, but it was much improved after 1885 when its symmetrical design was more apparent. No doubt as a gesture to the Italians (who had responded to news of *Inflexible*'s building by increasing the weight of *Duilio*'s guns to 90 tonnes (100 tons), it was despatched to join the Mediterranean Fleet in October 1881. Its guns were put to use at Alexandria in July 1882, in suppressing an Egyptian uprising against the British-backed Khedive, when it also took a few hits from shore forts. It remained in the Mediterranean until 1885, and after refitting and a few years spent largely in reserve, returned there in 1890–93. From 1893 to 1897 it served as guardship at Portsmouth, then went to Fleet Reserve in 1897, Dockyard Reserve in 1901, and was sold for scrapping in September 1903.

Modernisation

In the early 1880s, control positions and conning towers were still rudimentary structures. On *Inflexible* a control station was placed between the funnels, built over an air trunk coming up from below, though its position just above and between the guns was highly uncomfortable during firing. A small armoured box-like structure was also provided just forward of the leading funnel. Communication between the different stations on board a ship would soon be improved by the use of the telephone rather than the up-and-down voice-pipe, or the running boy-messenger. But at this time, much reliance was still placed on individual gun-laying. The accuracy and regularity of fire were poor, and increasingly worse at any range beyond 914m (1000yd), especially when the gun platform and the target were both moving, on different courses. New aids to gunnery were needed – an issue that would be addressed in the next decade.

Collingwood (1887)

Breech-loading guns mounted in barbettes fore and aft, secondary guns grouped in lateral batteries, a steaming speed of 16 knots, and big guns mounted higher than ever before in a British ship, were the prime features of HMS *Collingwood.*

In the late 1870s the French Navy introduced the *Formidable* class of battleship, which carried three large-calibre breech-loaders on the centre-line and six 140mm (5.5in) guns on each side of the main deck (it was not actually completed until 1889), and the British Admiralty, already aware of Italy's large *Italia*, was reviewing its options for new designs. A low-slung turret ship was not the answer: what was needed was a ship with high freeboard capable of long-range cruising.

The eventual answer was supplied in HMS *Collingwood*, laid down at the navy yard in Pembroke on 12 July 1880, launched on 22 November 1882, and completed in July 1887. The cost was £636,996 and it was Great Britain's first 'barbette battleship',

Flooded sections
As an experiment, the unarmoured fore and aft sections were flooded, with a loss in speed of only half a knot.

Torpedo nets
Torpedo nets, hung out at waterline level from lateral booms, were fitted to all major ships and used when lying at anchor.

Belt
Though it was termed a 'citadel ship', the armoured area in fact was a belt 42.7m (140ft) long and 2.3m (7ft 6in) high protecting the barbettes and machinery.

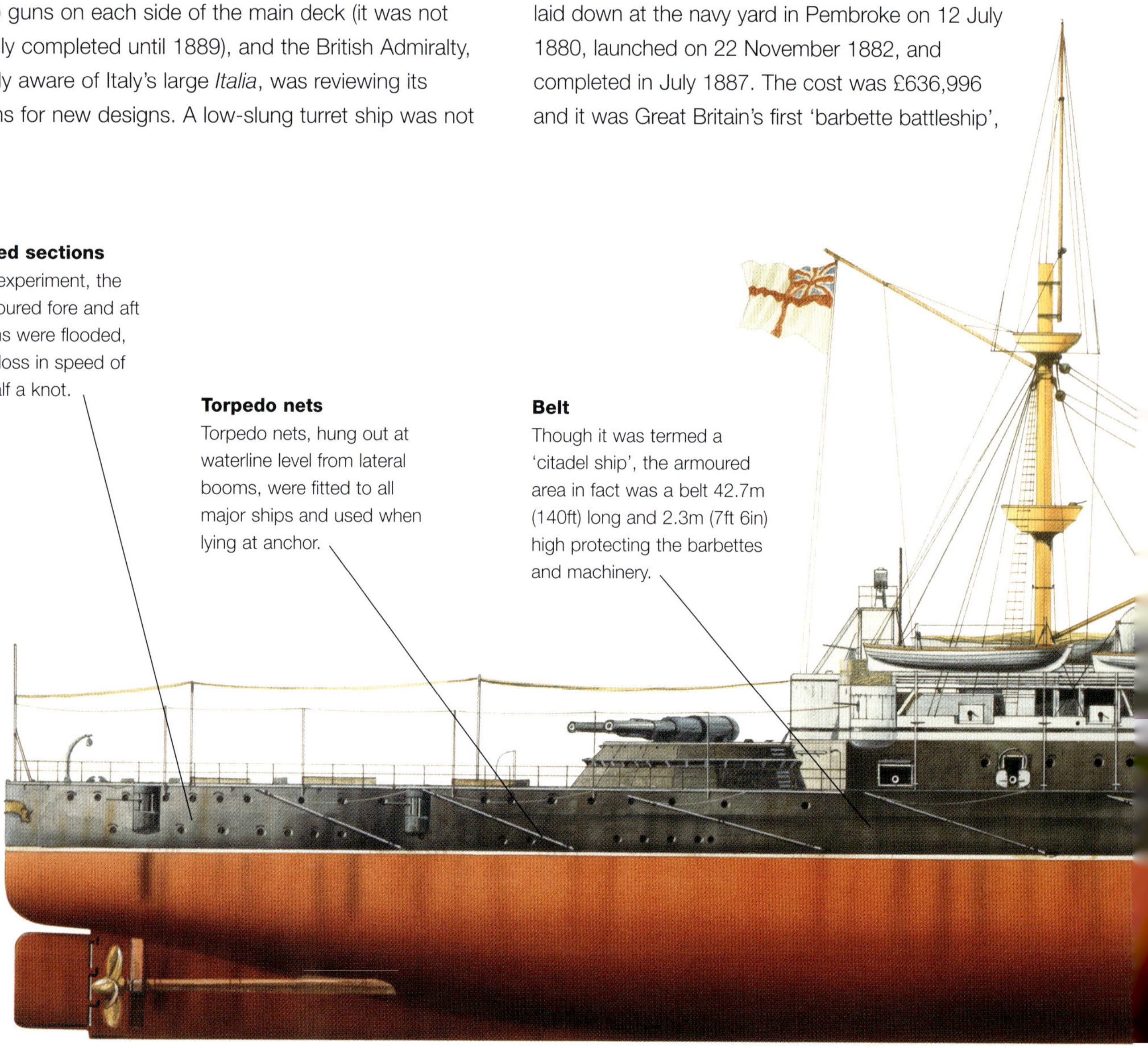

with its main guns set in armoured barbettes raised above the open deck fore and aft of the superstructure. A barbette could mount the gun higher, and *Collingwood*'s main guns were 8.5m (28ft) above the waterline, keeping drier in a sea, and with a better position for plunging fire. It gave less protection than a turret, but many gun crews disliked the restricted space and foul air of a turret, and the open position of the gun improved sighting and laying. The increasing number of torpedo boats in all navies also made it necessary to install a substantial secondary armament, and to give a good lookout for low-lying craft and torpedo trails, lofty platforms were built on the mainmast – now sometimes the only mast, as in *Collingwood*.

By this time, the convention of having all accommodation below the weather deck was being dropped. The weight of an iron hull and large engines made it feasible to build a superstructure above the deck as well as to place heavy guns at a high level.

Forced draught
This was the first RN ship to employ forced draught: injecting air into the 36 furnaces at above atmospheric pressure, to increase combustion and steam production.

The turret

Gun emplacement was passing through a phase of development. *Collingwood*'s heavy guns were mounted in open barbettes, which were fixed, armoured positions, with a trunk opening through the armoured floor to the magazines. The gun revolved within it, on its own turntable. Protection for the gun crew was limited to the height of the barbette's rim. Battleship design would move towards the turret, which provided a complete armoured covering for the gun and crew, and revolved with the gun. The barbette became the foundation of the turret, still with the trunk and hydraulic or electric hoisting systems from the magazines. The turret had some initial disadvantages, in that it added weight, could fill with fumes which were difficult to disperse, and made sighting more difficult. Fans and periscopic sights, and centralised gun direction, reduced these problems, while armoured turrets gave the ship greater staying power in a battle. Light guns had little or no protection until the twentieth century.

Storage and firepower

Side armour was concentrated in short amidships lines extending 1.5m (5ft) below the waterline and half that distance above, linked by transverse bulkheads. The bulkheads and the sides were lined with coal bunkers. In the unarmoured sections of the hull, the space between the armoured deck and the main deck was divided into compartments holding further fuel supplies and stores. Ammunition was stored beneath the armoured deck, with a heavily protected ammunition trunk, 3m (10ft) wide, passing through that deck and through the armoured floor of the barbette. The gun breeches were positioned above this trunk. The 152mm (6in) guns were set in a group, firing through gunports. Four of the 6-pounders were placed at the corners of this battery section, with eight on the boat deck above. Small sponsons extended from the after deck for light quick-firing guns, and others were placed in the fighting-top and on the superstructure.

Speed

With great emphasis being set on speed, the engines had forced draught applied. The 12 boilers worked at 90psi pressure. On trials, and so not under full load, it made 16.6 knots at an indicated horsepower of 6241kW (8369hp); the application of forced draught raised the output to 7139kW (9573hp) but added only 0.24 knots to the speed. Despite the comparatively high placing of the main guns, *Collingwood* had a low freeboard, and in a choppy sea the bow frequently submerged itself, making life uncomfortable for the crew, who inhabited the forecastle in the time-honoured way. But criticism was less focused on its qualities as a sea-boat than on its perceived lack of armour. The narrow waterline belt was criticised as inadequate.

The designers had quite deliberately set their sights (like Brin in Italy) on making *Collingwood* a fighting ship, and had taken great care in the plans for the unarmoured end, so that these could be riddled with shot and yet only

The side-to-side superstructure gives a quite different plan to the clear decks of earlier ships with centrally-mounted guns. Torpedo ports were at the corners of the superstructure.

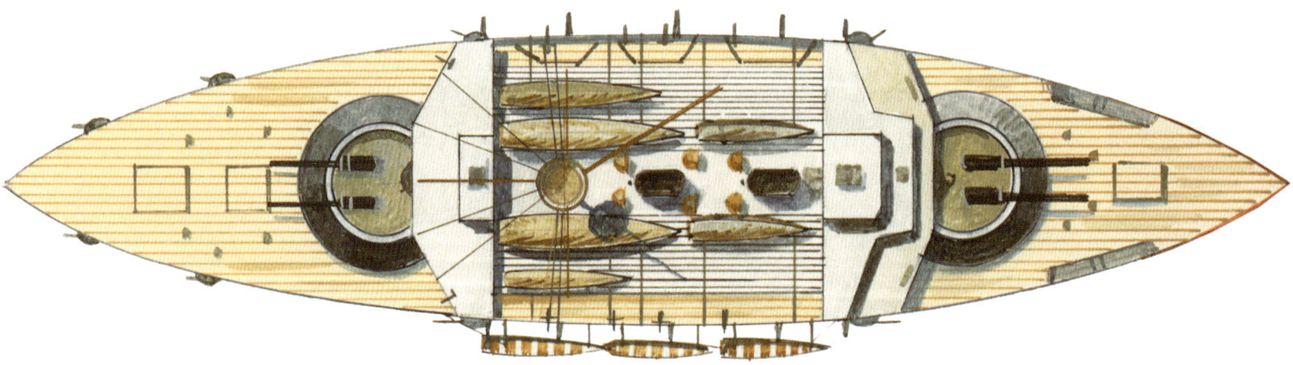

This illustration of *Collingwood* reveals the ship's lack of freeboard. Even in a modest sea, or at speed, the deck forward of the barbette would be awash. Note the capstan and anchor hoist.

take on water to a limited extent, and without impairing the crew's ability to fight. In addition, *Collingwood*'s guns could fire nine rounds a minute by comparison with the *Inflexible*'s two, making it a much more effective ship in battle and greatly increasing its chances of disabling an enemy vessel before incurring serious damage itself.

In reserve

In fact the ship's career did not give it any chances to prove its qualities in battle. Briefly commissioned for the Queen's Jubilee Review in July 1887, it was immediately put in reserve. Its longest stint of duty was nine years with the Mediterranean Fleet from November 1889 to March 1897, including a refit at the Malta dockyard in 1896. It was coastguard ship at Bantry from March 1897 to June 1903, then returned to reserve until 11 March 1909, when it was sold for scrap.

Specification

Dimensions	Length 99m (325ft), Beam 20.7m (68ft), Draught 8.03m (26ft 4in), Displacement 8618 tonnes (9500 tons)
Propulsion	12 boilers, 2 sets of Humphreys inverted compound 3-cylinder triple expansion engines
Armament	4 305mm (12in) 41-tonne (45 tons) BL guns, 6 152mm (6in) BL, 12 6-pounder, 14 small quick-firing guns, 4 356mm (14in) torpedo tubes (deck-mounted)
Armour	Belt 457–203mm (18–8in), Barbettes 292–250mm (11.5–10in), Conning tower 305–50mm (12–2in); Deck 76–50mm (3–2in)
Range	12,964km (7000nm) at 10 knots
Speed	16.8 knots
Complement	498

Royal Sovereign (1892)

This was the first British battleship to exceed 10,886 tonnes (12,000 tons) displacement, and the first to have steel-plate armour. In the words of one expert commentator, it was the first of a class of seven which 'sat the water with majesty and distinction'.

It took three-plus years from the laying-down to the completion of a first-class battleship. At the end of the 1880s British naval design was emerging from a period of often confused experiment and design which tried to catch up with what other nations were doing, into a more confident phase, in which a whole class of capital ships could be planned for. Shortly before that, there was a long moment when it seemed that the age of the battleship had already come to an end.

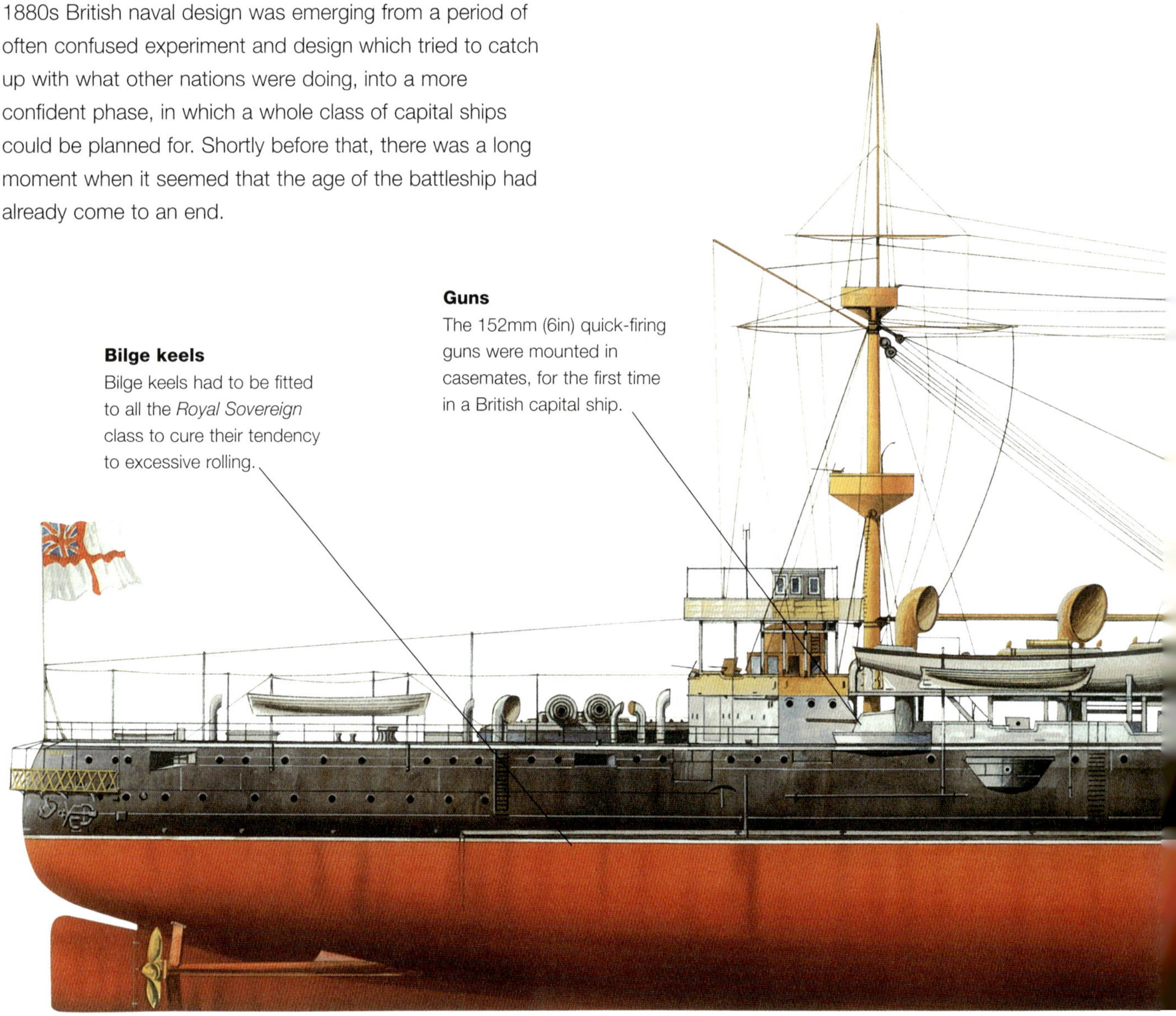

As the cost of a capital ship edged towards £1,000,000, politicians and planners were preoccupied by the fear that a single torpedo could potentially destroy it. France actually put a temporary stop to battleship construction in the late 1880s. But as it turned out, in the words of the First Lord of

This was the first class of British battleships to carry all their main armament on the weather deck, an innovation made possible by the 5.5m (18ft) freeboard and the use of barbettes rather than turrets.

Anchors
Double anchors were on the starboard only, not yet secured in the hawse-holes but with embrasures at deck level to hold them.

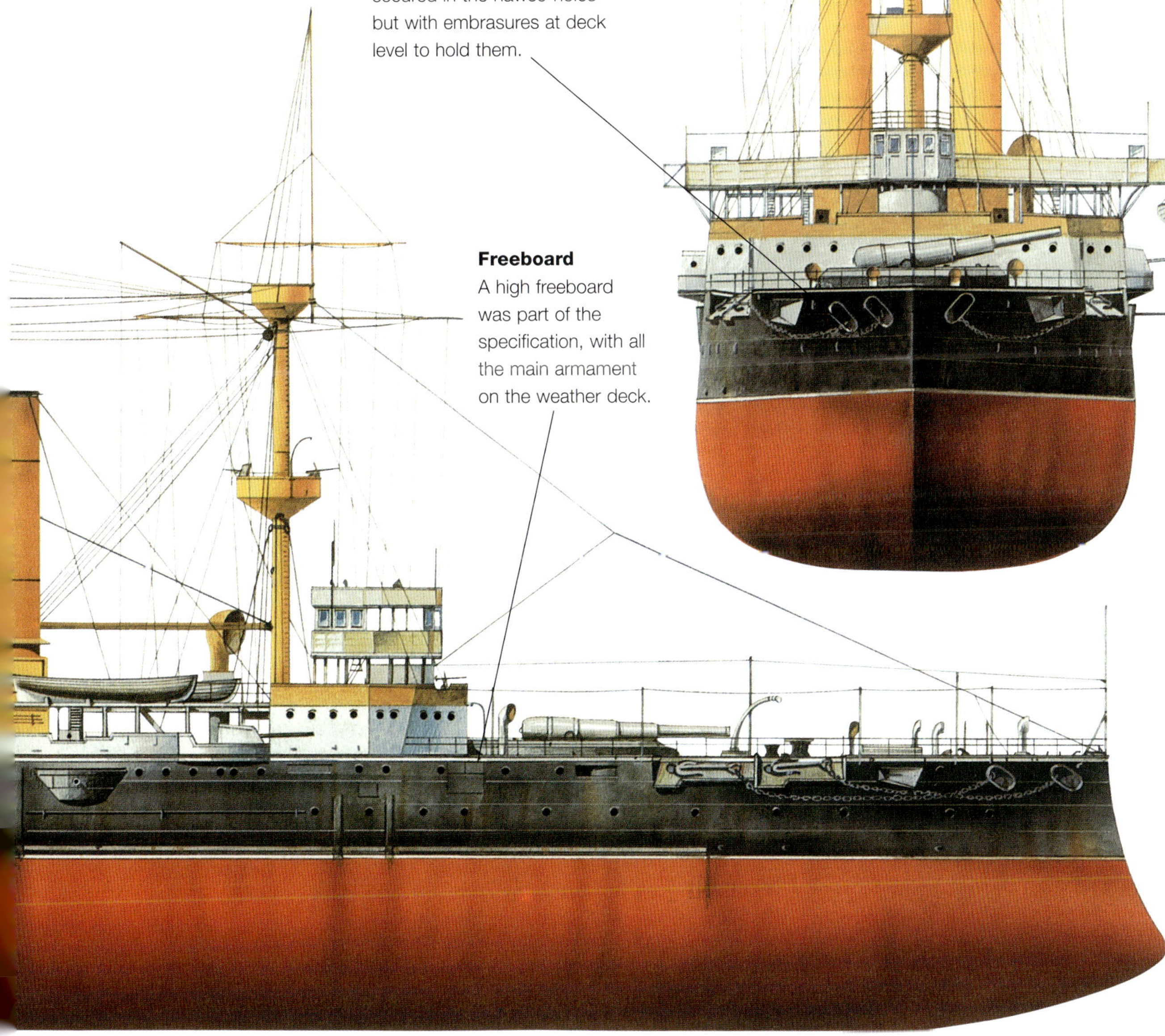

Freeboard
A high freeboard was part of the specification, with all the main armament on the weather deck.

the Admiralty in March 1889: 'the power of these torpedo boats had been greatly exaggerated by naval officers'. Capital ships still had a role. The Naval Defence Act of 1889 provided for the construction of no less than 70 ships, including eight large battleships of 12,836 tonnes (14,150 tons) and one of 9525 tonnes (10,500 tons). At the time, Britain possessed 22 first-class battleships and 15 second-class; France had 14 and 7 respectively, and Russia had 7 and 1. Germany, watching developments with interest, was about to expand its battleship strength.

Royal Sovereign class

The seven large battleships were to form the *Royal Sovereign* class, and the name ship was laid down at Portsmouth on 30 September 1889, launched on 26 February 1891, and completed in March 1892, at a cost of £913,986. It was a barbette rather than a turret ship and the consequent reduction in weight enabled it to ride higher in the water and to achieve a greater speed. This was quite a radical departure in design, compared with previous years, but familiar to those who recalled the sailing men of war; like those, *Royal Sovereign* had an inward slope of 'tumblehome' on the sides to help keep the vertical centre of gravity safely low (the Director of Naval Construction, Sir William White, is said to have copied contemporary French warships in this respect, though he did not incorporate sponsons, in the *Royal Sovereign* class anyway).

One member of the class, HMS *Hood*, was built as a turret ship and was conspicuously less successful than the other six, both as a seaboat and as a fighting ship, its guns sitting 1.83m (6ft) lower and its decks regularly awash. Like some preceding capital ships, including HMS *Trafalgar*, *Royal Sovereign* had two funnels side by side, placed above the division between the two boiler rooms which were set in line, with the engine room aft. As was by now standard, twin screws were fitted.

Armour

From the late 1870s, steel armour had been tried out, and in Britain a compound, with a steel face welded on to iron plates, had been in use from the mid-1880s. The French firm of Schneider had developed an effective steel armour by 1881, and both the Americans and the Russians used this. The British, unable to match the Schneider product, continued with the steel-faced compound. In 1889 a nickel-steel alloy was developed, and in 1891 the American metallurgist H.A. Harvey produced a new kind of 'Harvey steel', face-hardened by the application of carbon at very high temperature and subsequent tempering and annealing. *Royal Sovereign*'s belt and barbette armour was of the old compound type, but the hull parts beyond the belt were of 127mm (5in) 'Harveyised' steel, backed by coal bunkers. This was considered to offer enough resistance to medium-calibre shells.

Specification

Dimensions	Length 115.8m (380ft), Beam 22.86m (75ft), Draught 8.5m (28ft) (full load), Displacement 14,138 tonnes (15,585 tons) (full load)
Propulsion	8 cylindrical single-ended boilers, 2 sets of 3-cylinder vertical triple expansion engines, 6711kW (9000hp), twin screws
Armament	4 343mm (13.5in) guns, 10 152mm (6in) QF, 16 6-pounder QF and 12 3-pounder QF guns; 7 356mm (14in) torpedo tubes, 2 submerged
Armour	Belt 457–356mm (18–14in), Bulkheads 406–356mm (16–14in), Deck 76–63mm (3–2.5in), Barbettes 432–279mm (17–11in), Casemates 152mm (6in), Conning tower (forward) 356mm (14in), aft 76mm (3in)
Range	8741km (4720nm) at 10 knots
Speed	18 knots (maximum under forced draught)
Complement	712

Firepower

Royal Sovereign's barbettes, housing four 343mm (13.5in) BL guns, were of similar shape and design to those of *Collingwood*, of 152mm (6in) armoured steel, though the armoured walls extended down to belt level. The secondary

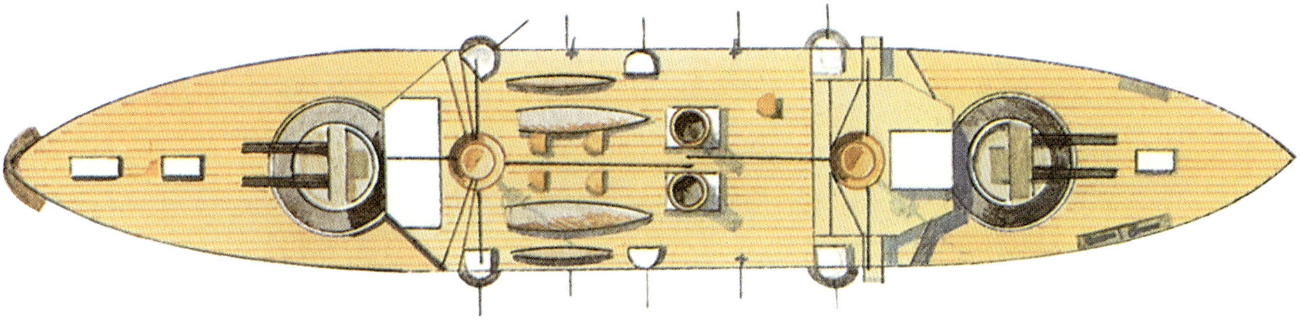

The barbettes were pear-shaped in plan, extending under the deck structures. Though with a maximum 432mm (17in) armour depth, their exposure of the guns was a weakness.

armament of 10 152mm (6in) guns were placed in armoured casemates 152mm (6in) thick on the main deck level, and behind gun-shields on the upper deck (altered to casemates on refit in 1903). Six-pounder guns were distributed along both decks and 3-pounders were placed on the shelter deck and in fighting tops of both masts. Seven 356mm (14in) torpedo tubes were incorporated, but this was altered to four 456mm (18in) in a 1903–04 refit at Portsmouth. *Royal Sovereign* was initially assigned to the Channel Fleet and was part of the British squadron at the ceremonial opening of the Kiel Canal in June 1895. From June 1897 to August 1902 it served in the Mediterranean (one of the 152mm (6in) guns exploded in November 1901, killing six men) and was then guardship at Portsmouth until 1905.

On reserve at Devonport from May 1905 to February 1907, it was subsequently put on 'Special Service' with a skeleton crew. From April 1909 to 1913 it was a part of the 4th Division of the Home Fleet. In October 1913 it was sold off.

The marine artist William Frederick Mitchell (1845-1914) was commissioned to paint a range of British warships. This chromolithograph shows *Royal Sovereign* cruising, with guns slightly raised.

Maine (1895)

Despatched to Havana, Cuba, at a time of high international tension to 'show the flag' and protect US interests, *Maine* blew up in an explosion whose cause would remain a subject of controversy for more than three quarters of a century.

Maine was authorised by Congress on 3 August 1886 and laid down in the New York Navy Yard as an armoured cruiser on 17 October 1888, but was redesignated as a battleship of the second class and launched on 18 November 1890. It was not commissioned until 17 September 1895. Construction cost was $4,677,788. Battleships were a relatively new thing in the US Navy and *Maine* was only the second in the pre-*Dreadnought* era.

Though the USA had pioneered the ironclad ship in the 1860s, these had not been large ocean-going craft capable of long cruises. British and European battleship design was drawn on for *Maine*.

Coal bunkers
Coal bunkers were placed between the forward 254mm (10in) magazine and the outer hull. Spontaneous combustion here may have caused the fatal explosion.

Armament
152mm (6in) guns were fitted in casemates at the bow and stern, to give supplementary fore and aft fire.

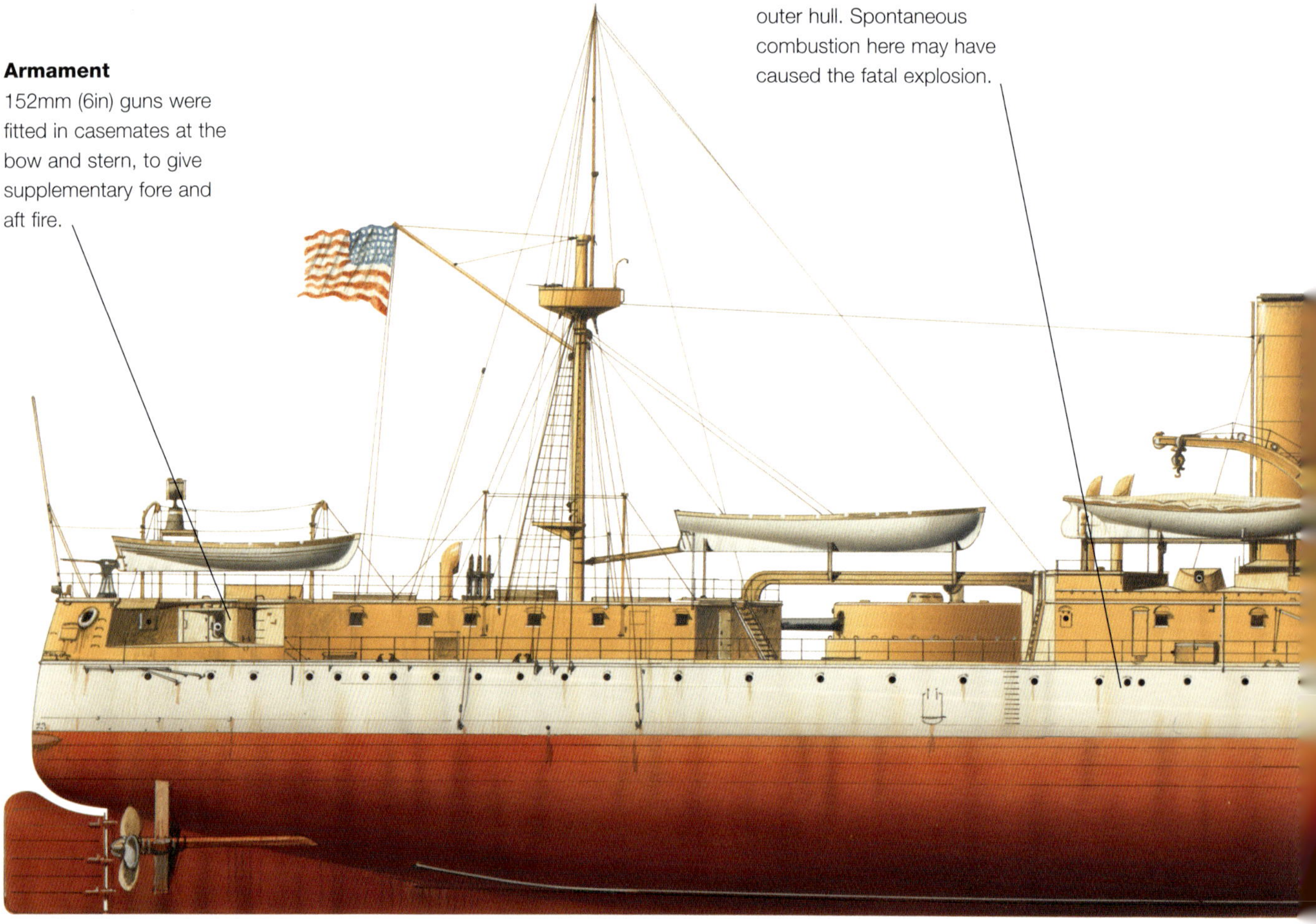

Maine

Strong elements of US public opinion were hostile to the idea of long-range capital ships, and the battleships of the 1890s were officially designated as coastal, emphasising a defensive role. At that time several South American countries were expanding their navies, and Brazil had recently acquired two British-built battleships, *Riachuelo* and *Aquidaba*. The United States government felt it was essential to match these developments. The Panama Canal was not yet open (completed 1914) and the USA had to maintain separate fleets on each coast.

Turrets
The two main turrets were not counterbalanced, and if both were trained in one direction the ship heeled. After *Maine* and *Texas* the echelon formation was dropped.

Fighting tops
Each fighting top held a 1-pounder gun, part of the anti-torpedo boat defences (note also the searchlights). These guns fired .5kg (1.1lb) shells to a range of approximately 3200m (3500yd).

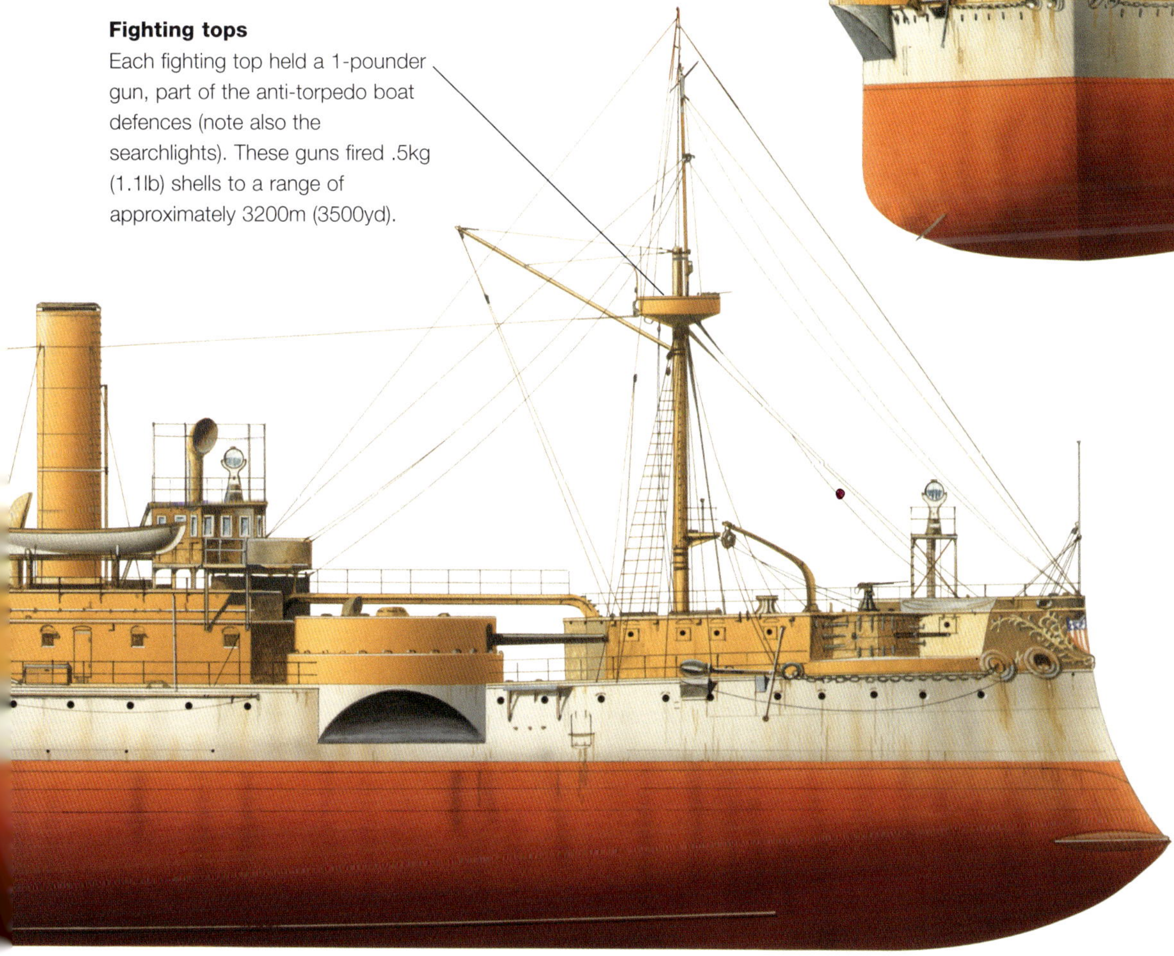

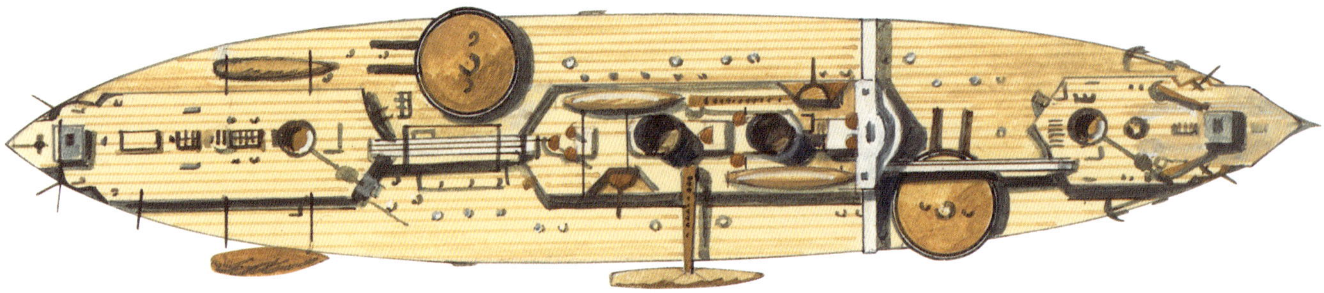

Deck plan: the echelon placing of the gun turrets was made necessary by the raised forecastle and poop decks, an aspect of design which would disappear in future USN battleships.

Slow start

Maine's slow completion was due to difficulty in securing supplies of suitable armour plate, and consequently when finally commissioned, it was already a ship of an earlier generation, built to concepts and standards of the early 1880s. Its two gun turrets, each with two 254mm (10in) guns, were set in an elongated echelon formation, extending over the sides on sponsons, with the starboard one set well forward in front of the superstructure. The shells weighed 236kg (520lb) and with the gun set to maximum elevation of 15 degrees, had a range of 18,000m (20,000yd). As on similar European ships, broad deck spaces were kept clear to allow for the possibility of firing directly forward and aft, though as with other ships, this was not in fact practicable because of the risk of blast damage. The turrets revolved on hydraulic power and could be loaded at any angle of train. The secondary armament of six 152mm (6in) guns was set in casemates on each side. Six-pounder guns, intended as anti-torpedo boat defence, were placed higher on the superstructure and one was installed in each fighting top. Four 457mm (18in) torpedo tubes were fitted, all on the main deck. The original design also provided for two single-tube torpedo boats to be carried, but these were not mounted.

Maine was also endowed with a ram bow, though the steel construction of the hull was unsuitable for ramming tactics. Within the hull there were 214 watertight compartments. A watertight bulkhead running lengthwise separated the two engine compartments, and a double bottom underlay the central armoured box which, because of the widely spaced turrets, accounted for almost two thirds of the ship's length.

Eight single-ended marine boilers, operating at 135psi, supplied steam to two inverted triple expansion engines, mounted vertically. Two dynamos were fitted for electric lighting and searchlights. The engines were to have been supplemented by sails on three masts with a barque rig, but between launch and completion the third mast was removed and the others fitted as 'military' masts with fighting tops and signal yards.

Specification

Dimensions	Length 98.9m (324ft 4in), Beam 17.4m (57ft), Draught 6.5m (21ft 5in), Displacement 6062 tonnes (6682 tons)
Propulsion	8 boilers, 2 N.F. Palmer inverted vertical triple expansion engines, 6930kW (9293hp), 2 screws
Armament	4 254mm (10in) guns, 6 152mm (6in) guns, 7 6-pounder and 8 1-pounder guns; 4 457mm (18in) torpedo tubes
Armour	Belt 305mm (12in), Turrets 203mm (8in), Conning tower 254mm (10in), Deck 102mm (4in)
Range	Not known
Speed	16.45 knots
Complement	374

Explosion

Cuba in 1897 was still a Spanish colony but an independence struggle was underway, and in line with long-established American policy against European involvement in American waters, *Maine* was despatched to Havana as a deliberate show of force, arriving on 25 January 1898, anchoring in the harbour. On 15 February at 21:40, there was a massive explosion in the forward part of the hull and the ship was left a half-submerged wreck, with 252 crew dead and missing. Given the tensions of the time, an attack could not be ruled out and a US court of inquiry decided that *Maine* had been blown up by a mine, though how, or by whom, was not clear. The incident was a major factor in hastening the Spanish-American War, which began on 21 April. Over the years subsequent inquiries and many different theories explored the cause of *Maine*'s destruction. In 1975 Admiral Hyman Rickover's report concluded that the explosion was an internal one, most probably caused by a fire in the coal bunker adjoining the reserve magazine of 152mm (6in) ammunition, and that remains the likeliest explanation. The remains were raised and towed out to be sunk in the open sea with full naval honours on 12 March 1914.

The part-submerged wreck of *Maine* lay in Havana harbour from 1898 to 1914.

Spanish-American War

Maine's destruction hastened the impending war between the USA and Spain. Its two naval engagements were American victories. The Battle of Manila Bay, 1 May 1898, was fought between cruisers and the Battle of Santiago de Cuba on 3 July 1898, fought between four US battleships, one armoured cruiser, and two auxiliary cruisers, and four Spanish cruisers and two destroyers. The battleships were *Texas*, *Iowa*, *Oregon* and *Indiana*. *Oregon* had been summoned from the Pacific Fleet and made the traverse via Cape Horn in 67 days. During the voyage one of its bunkers caught fire, as may have happened on *Maine*, but the fire was put out. The Spanish ships were seeking to escape from the blockaded harbour of Santiago rather than to confront a superior enemy force, but the Americans manoeuvred to prevent this. All six Spanish ships were sunk, or run aground by their crews. Although the American ships were hit many times, they suffered no major damage and only one man was killed.

Indiana (1895)

A first-class battleship described by a British expert as 'distinctly superior to any European vessels of the same displacement', *Indiana* and its two sister ships were classified as 'coastal battleships' partly for domestic political reasons.

The influential American naval strategist Captain Alfred Thayer Mahan considered that defence of the American coastline needed a 'forward strategy' to intercept an enemy fleet out at sea. This required battleships, and Mahan's view was endorsed by a policy committee in July 1889, which recommended construction of a battlefleet. Authorised by Congress in 1890 and recorded as BB-1 in the US Navy's list of capital ships, *Indiana* was laid down at Cramp's yard in Philadelphia on 7 May 1891, launched on 28 February 1893, and commissioned on 20 November 1895. Total cost was almost $6,000,000.

As with USS *Maine*, completion was delayed by the slow delivery of armour plate, and preliminary sea trials were conducted before the armour was applied or guns mounted.

Protection was intended to equal the British *Royal Sovereign* class, but the availability of Harvey steel gave lighter weight but more effective protection to the belt and main turrets; other armour was nickel steel. With two tall funnels and a high single mast, *Indiana* presented quite a lofty appearance, accentuated by a low freeboard.

***Indiana* was classified as Battleship No. 1, and sister ships *Massachusetts* and *Oregon* were Battleships 2 and 3. From 1907 the designation was sometimes expressed as B-1, etc. The initials BB were used from 17 July 1920.**

Firepower

Naval gunnery entered a phase of rapid development in the 1890s, with a strong emphasis on improving the precision and rapidity of fire. The US and British navies were leaders in these respects. *Indiana* was heavily armed for a ship of its size, and gun-positioning was a problem. Critics saw this as a consequence of trying to pack too many guns into too little space. If the 203mm (8in) guns were trained more than 30 degrees forward or aft of the beam, the effect of the blast made the sighting hoods of the 330mm (13in) turrets untenable. The blast of the 203mm (8in) guns also made use of the 152mm (6in) guns impossible, and some of the 152mm (6in) gun-ports had to be closed up (in 1911 the 152mm (6in) guns were replaced by 76mm (3in) guns). In addition, a wide range of ammunition had to be carried. Though American ships were perhaps extreme examples, this was a feature of all pre-Dreadnoughts.

Guns

Heavily-gunned for their size, *Indiana* and its class-mates *Oregon* and *Massachusetts* carried 330mm (13in), 203mm (8in) and 152mm (6in) guns. The main guns were mounted in pill-box turrets which were not counterbalanced, making the ship tilt when the guns were trained abeam; and making the turrets themselves unstable. The problem was accentuated by the ship's tendency to roll, which was partially cured by the fitting of bilge keels, but the low freeboard meant that even in a moderate sea the decks were awash, especially when travelling at speed. The 203mm (8in) semi-heavy guns were intended to pierce the thickened side armour which had been introduced to counter the effect of explosive shells from quick-firing 152mm (6in) guns, but their weight and the blast effect of their discharge on a relatively short hull reduced their effectiveness in use.

Specification

Indiana's original specification turned out to be unrealistic in certain ways, once the ship was in service. Coal capacity had been reckoned at 363 tonnes (400 tons) whereas in normal service it carried more like 1451

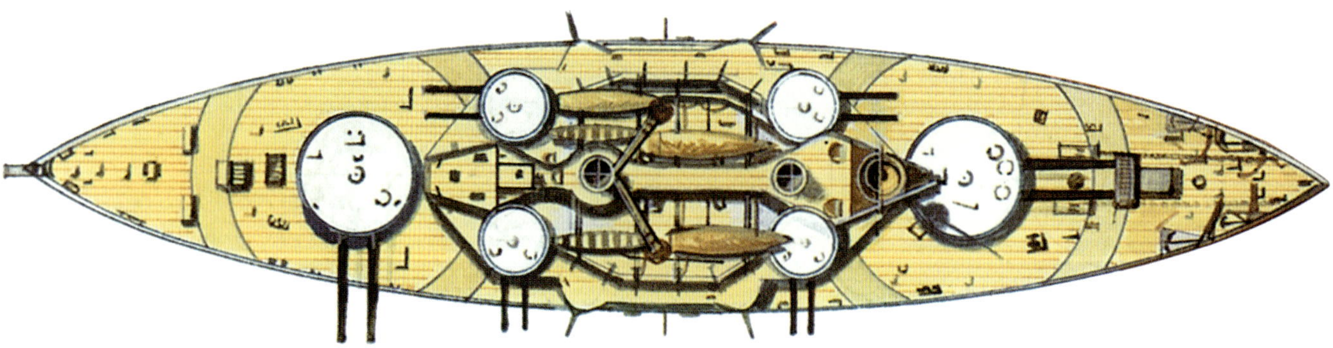

The heaviest guns are now on the centre-line, but the eight 203mm (8in) guns mounted at the corners of the central armoured area brought criticisms of 'over-gunning' on the relatively small US pre-Dreadnoughts.

Indiana in a side-on view showing her four 330mm (13in) heavy gun turrets fore and aft.

tonnes (1600 tons). The design had provided for a draught of 7.3m (24ft), but on completion the weight of equipment had increased it to 7.8m (25ft 8in); and fully loaded with coal, stores and ammunition the draught was 8.2m (27ft), which put the whole armoured belt below the waterline, as well as lowering the freeboard to make it a very 'wet' ship. Later battleship designs would be required to allow for a normal load of at least two-thirds of maximum capacity.

Indiana was deployed to the North Atlantic Squadron based at Key West and saw action in the Spanish-American War of 1898, shelling shore targets and working on interception missions against Spanish movements. *Oregon*, built at San Francisco and intended for Pacific service, was sent round by Cape Horn to join the Atlantic fleet, then returned to the Pacific.

On 3 July *Indiana* was part of the US force which sank the Spanish destroyers *Pluton* and *Furor*, though engine problems prevented it from chasing two Spanish cruisers, which escaped.

Specification

Dimensions	Length 106.96m (350ft 11in), Beam 21.11m (69ft 3in), Draught 8.2m (27ft), Displacement 9333 tonnes (10,288 tons)
Propulsion	4 double-ended boilers, 2 inverted vertical triple-expansion engines developing 7295kW (9738hp), 2 screws
Armament	4 330mm (13in) guns, 8 203mm (8in) guns, 4 152mm (6in) guns, 20 6-pounder, 6 1-pounder guns; 4 457mm (18in) torpedo tubes
Armour	Belt 460–220mm (18–8.5in), Main turrets 380mm (15in), Conning tower 254mm (10in), turrets 152mm (6in), Deck 76mm (3in)
Range	9100km (4900nm) at 10 knots
Speed	15.6 knots
Complement	473

Refit

From May 1900 to March 1901 it was on reserve, then used for training cruises. In a refit at New York Navy Yard between December 1903 and January 1906, eight new boilers replaced the original four, and the main turrets were balanced and had electric traversing gear installed. From 1905 to 1914 it was used as a training ship. During this period various further modernisations were made, including the fitting of the lattice 'cage' mainmast that would be typical of the US battleship. The 152mm (6in) guns were replaced by 76mm (3in), and radio communication was installed.

Indiana crossed the Atlantic to Queenstown (Cobh), Ireland in 1911. Again put on reserve at Philadelphia in 1914, when the USA entered World War I in 1917, it was recommissioned on 24 May and again used for gunnery training. Decommissioned in January 1919, the name was removed and as 'Coast Battleship No. 1' it was used as a target to assess the effect of aerial bombs on an armoured warship. Sunk in shallow water in 1920, the remains were retrieved and sold for scrap in 1924.

Jauréguiberry (1896)

Battleships were controversial in the French Navy, with some strategists and naval architects arguing that they were unnecessary and even outmoded as ships of war. The *Jauréguiberry* class were the first French battleships to have guns mounted in turrets rather than in barbettes.

Named for a famous naval commander, Bernard Jauréguiberry, the ship was officially classified as a *cuirassé d'escadres* (fleet armoured ship). Laid down at La Seyne-sur-Mer on 23 April 1891, it was constructed to plans by the naval architect Amable Lagane, launched on 27 October 1893 and entered service on 30 January 1896. Although four other ships formed a class with it, their appearance was different in each case, and all they had in common was the main armament. Known as the 'fleet of one-offs', they were not a particularly successful set of ships, with another shared factor being their instability. *Jauréguiberry* was the most intensively used of the five.

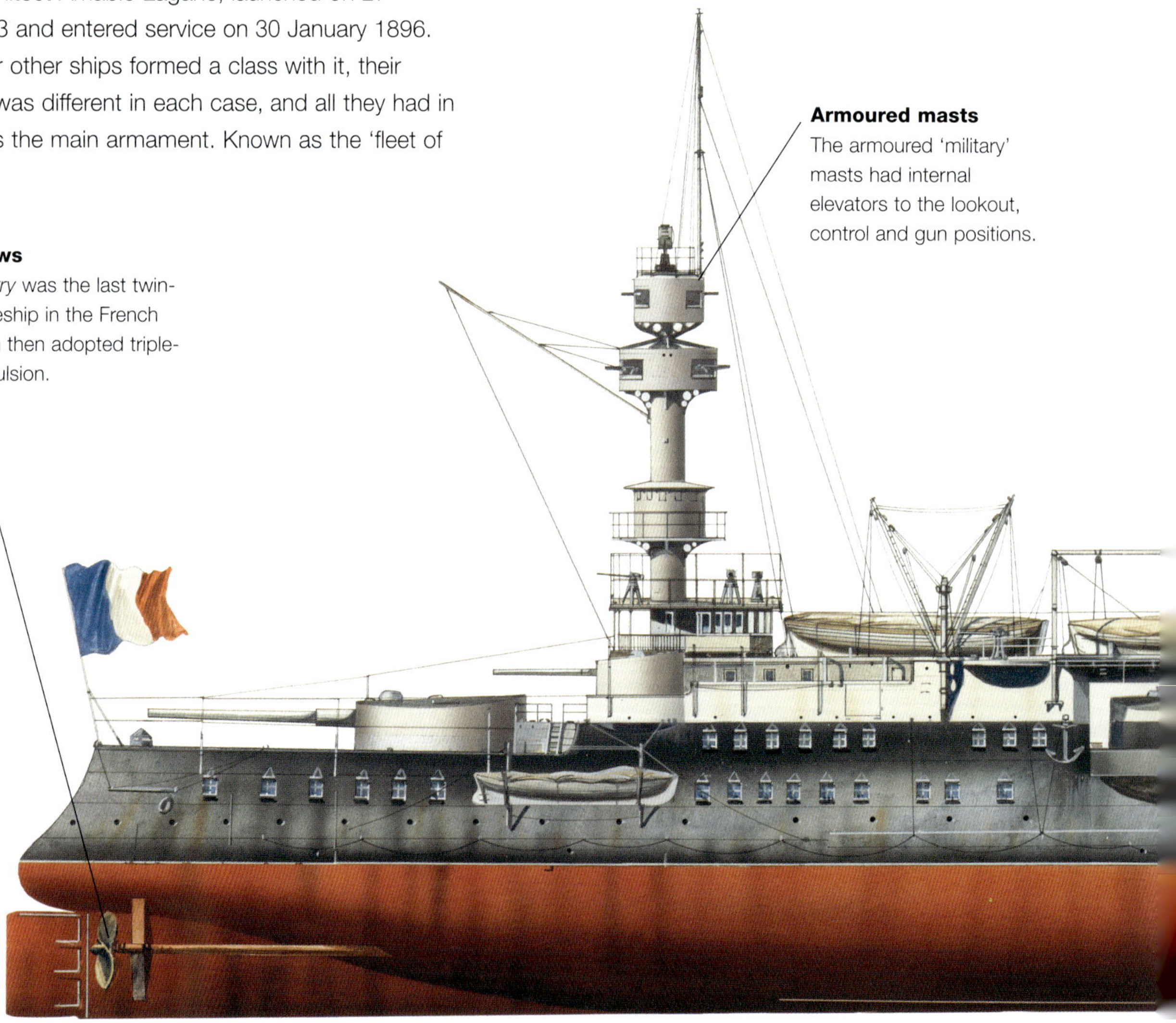

Armoured masts
The armoured 'military' masts had internal elevators to the lookout, control and gun positions.

Twin screws
Jauréguiberry was the last twin-screw battleship in the French Navy, which then adopted triple-screw propulsion.

French battleship design had a style all of its own, backed up by intensive work and study on ship behaviour, though this was often of a theoretical rather than practical kind.

Masts

The relatively massive constructions on the masts would have made vulnerable targets in a gunnery action.

Guns

Gun mountings were electrically operated and had all-round loading.

Boat launching gear

One of the many variations between it and the other ships in its class was the ingenious but complicated boat-launching gear, needed to clear the tumblehome sides.

Jauréguiberry

Facts and figures

Jauréguiberry was 7m (23ft) shorter than any of the other ships in the 'class', and the main guns were placed at extreme positions fore and aft. Its maximum beam was 23m (75ft 6in) but the bulging 'tumblehome' construction of the hull meant that the main deck was relatively narrow. Two massive sponsons just aft of the after funnel supported two 274mm (10.8in) guns. The ship was solidly armoured with Le Creusot nickel steel, applied in a waterline belt with an upper belt above. The armoured deck was placed at the upper level of the waterline belt.

The positioning of the big guns gave the ship a field of fire in all directions, with up to three able to fire a 'broadside'. The firing arc of the 305mm (12in) guns was 250 degrees, and at maximum elevation of 15 degrees could send a 340kg (750lb) shell 12,000m (13,000yd), which was rather more than the maximum range anticipated for ship-to-ship fighting in the years before 1910. Secondary armament was installed with anti-torpedo defence in mind, and consisted of eight 138mm (5.4in) guns in twin turrets placed at the four corners of the superstructure.

Two vast columnar masts with fighting tops and lookout posts rose above. With both funnels forward of the centre-line, the ship had a long-tailed look. Although Lagane was a highly gifted ship designer, the 'tumblehome' form was somewhat discredited after the sinking of Russian battleships of the similarly-hulled *Borodino* class at Tsushima in 1905, and it was not perpetuated.

Specification

Dimensions	Length 112.6m (377ft 4in), Beam 22.15m (72ft 6in), Draught 8.45m (27ft 9in) Displacement 10,919 tonnes (12,036 tons) full load
Propulsion	24 Lagrafel d'Allest watertube boilers, 2 vertical inverted triple expansion engines, giving 10,769kW (14,441hp), 2 screws
Armament	2 305mm (12in) guns in single turrets, 2 274mm (10.8in) guns in single turrets, 8 138.6mm (5.4in) guns in twin turrets, 8 100mm (3.9in) and 16 3-pounder guns, 4 457mm (18in) torpedo tubes
Armour	Waterline belt 400–160mm (15.7–6.3in), Upper belt 170–120mm (6.7–4.7in), Armoured deck 90mm (3.5in), Main turrets 370–280mm (15–11in), Conning tower 250mm (9.8in)
Range	7260km (3920nm) at 10 knots
Speed	17.07 knots
Complement	597

Disaster prone

Jauréguiberry's career was marked by a series of minor disasters. On 30 January 1896 trials began, but were held up by a burst boiler tube and damage to the firing mechanism of one of the 305mm (12in) guns and did not resume until January 1897; then in March of that year a torpedo's air chamber exploded, fortunately with only

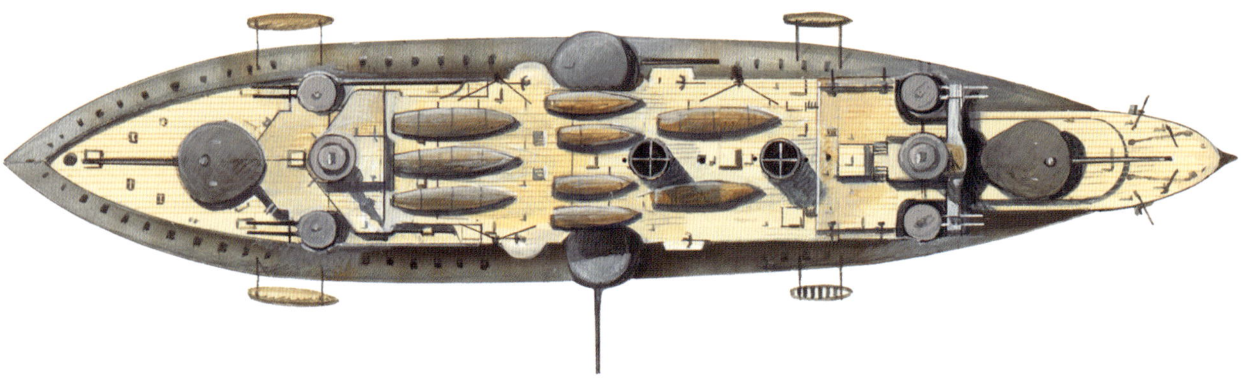

Deck plan: the *Jauréguiberry* carried eight gun turrets on a cramped deck space that included more than 30 guns of various sizes in total.

Jauréguiberry and its sisters were deliberately designed to look like floating steel fortresses. A consequence of this was greatly restricted deck space compared to other battleships.

minor damage, and it became the flagship of the Mediterranean fleet in May. In February 1904 it was transferred to Brest and the Northern Squadron, and was damaged after hitting a rock off Brest.

On a visit to Portsmouth in 1905 it collided with an English steamer and in the same year suffered damage to its propellers from a torpedo fired from the *Sagaie*. During repairs in 1906 the torpedo tubes were removed. In 1907 it was based at Toulon and placed in the reserve division of the Mediterranean Squadron until April 1908.

From then it remained in service alternately at Brest and Toulon, and in October 1913 became the flagship of the Training Division. In World War I, the oldest French battleship still on the active list, it went on service in the Mediterranean, initially as a troop carrier and escort, and from March to August 1915 was French flagship in the Gallipoli campaign, firing on shore fortresses (and receiving minor damage); then it was based at Port Said to defend the Suez Canal until decommissioned in 1917. Two of the 305mm (12in) guns were unshipped and left behind to provide canal defences. On 6 March 1919 *Jauréguiberry* returned to Toulon for disarming, and was struck from the list on 20 June 1920. It continued in use as an accommodation hulk for engineers at Toulon until 1932. In July 1934 it was sold for scrap.

New vessels

For around 10 years from the mid-1880s, no new battleships were built for the French Navy, due to the influence of the so-called *jeune école* (young school) of designers. In their view, the commerce-raiding cruiser was the vessel to concentrate on, rather than the line-of-battle ship, which they considered outmoded, too expensive, too slow and too vulnerable to torpedo attack. However, since the British, the Germans, the Russians and the Americans were continuing to build battleships, the French eventually resumed, and from 1895 a range of ships of distinctive and individual appearance appeared. Certain basic characteristics were common to all, including the lozenge arrangement of big guns set in single turrets, with the fore and aft ones placed very close to stem and stern, which was continued until 1903. All had massive ram bows, and heavy masts. *Jauréguiberry* had many up-to-date features when completed, including watertube boilers, but was in poor condition by 1914.

Kearsarge (1900)

Bearing an already-famous name, *Kearsarge* as a member of President Theodore Roosevelt's 'Great White Fleet' became one of the best-known of American warships. But it also had a valuable later incarnation.

The first *Kearsarge*, a screw sloop, sank the Confederate commerce raider *Alabama* in the American Civil War on 19 June 1864. The next, BB-5 on the Navy List, was laid down on 30 June 1896 at Newport News Shipbuilding Co., launched on 24 March 1898, and completed on 20 February 1900. It cost $1,849,380.

The freeboard was 0.9m (3ft) higher than *Indiana*, showing that some lessons were being learned, but its nominal coal capacity was still only 453 tonnes (500 tons). The side armour was thickened to 381mm (15in) and the armoured deck sloped downwards towards the ends from the top of the main belt.

A long central casemate carried seven 127mm (5in) American-built guns on each side: its 152mm (6in) 'Harveyised' steel armour was intended to resist 152mm (6in) enemy shells fired from 914m (1000yd). Splinter bulkheads of 50mm (2in) steel separated each gun compartment within the casemate.

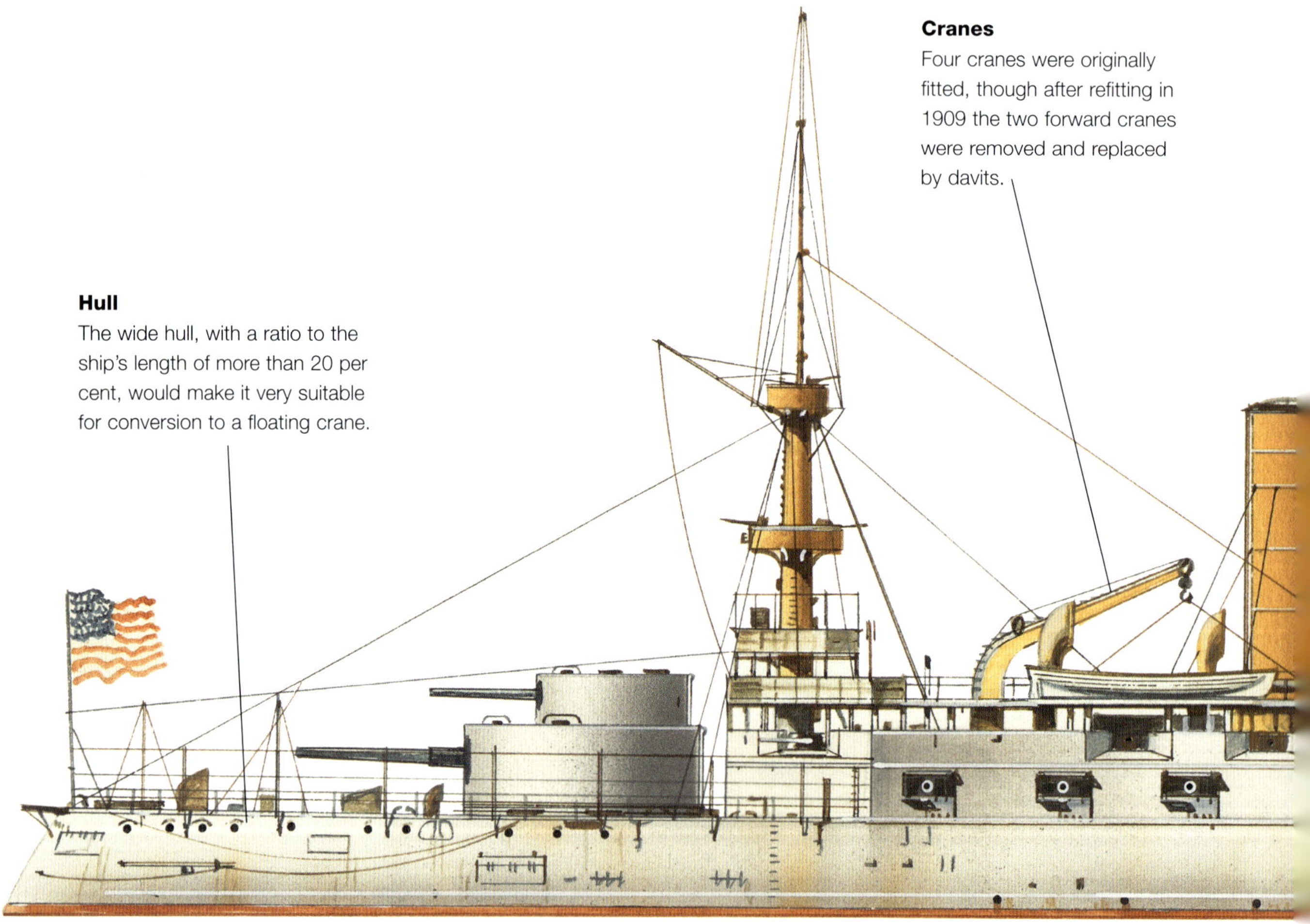

Cranes
Four cranes were originally fitted, though after refitting in 1909 the two forward cranes were removed and replaced by davits.

Hull
The wide hull, with a ratio to the ship's length of more than 20 per cent, would make it very suitable for conversion to a floating crane.

Kearsarge in its original appearance. The two double turrets, fore and aft, turned as one. The secondary battery, amidships, is prominent.

Casemates
The seven-a-side casemates were a distinctive feature. Splinter protection was provided between the gun compartments.

Masts
The lofty pole masts were replaced by basket masts during the 1909 refit.

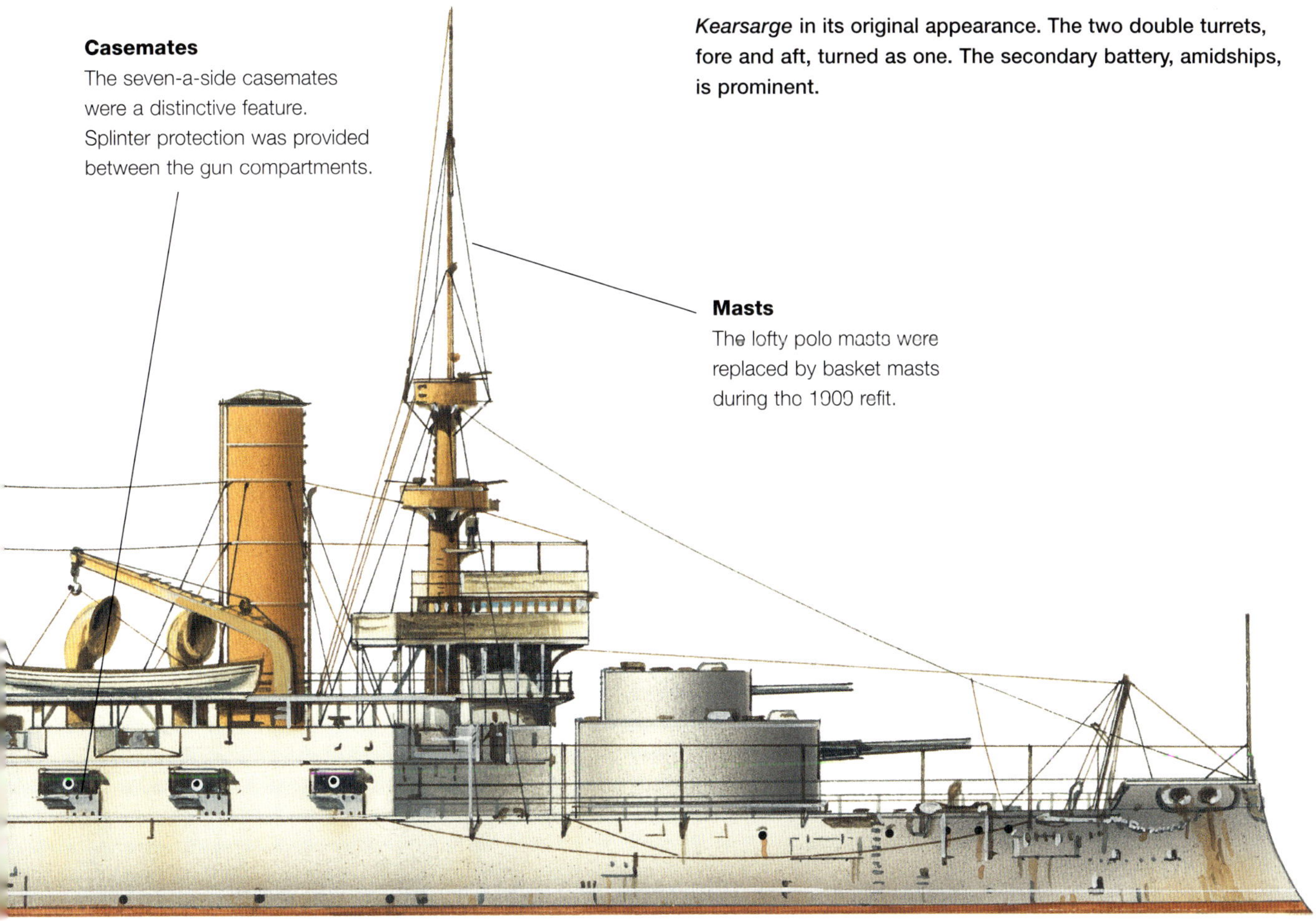

Specification

Dimensions	Length 114.4m (375ft 4in), Beam 22.02m (72ft 3in), Draught (as designed) 7.16m (23ft 6in), Displacement 10,469 tonnes (11,540 tons)
Propulsion	2 vertical triple expansion engines, 7457kW (10,000hp), 2 screws
Armament	4 330mm (13in) guns, 4 203mm (8in) guns, 4 152mm (6in) guns, 20 6-pounder and 8 1-pounder guns, 4 457mm (18in) torpedo tubes
Armour	Main belt 419–127mm (16.5–5in), Barbettes 381–318mm (15–12.5in), Main turrets 432-381mm (17–15in), Upper 279–15mm (11–6in), Conning tower 254mm (10in)
Range	5556km (3000nm) at 10 knots
Speed	16 knots
Complement	553

Innovations

Among the ship's innovations was 102mm (4in) bow armour. Another was the mounting of a powerful secondary armament, consisting of four 203mm (8in) guns. Their turrets were placed directly on top of the 330mm (13in) main turrets. While it resolved the problem of where to put them, it meant that both turrets could not fire at the same time. The advantages were a shared barbette and hoisting trunk, and control of both turrets by a single officer. But the disadvantages included heavy weight on the main turret bearings and the likelihood of all forward guns being put out of action by one hit. Though light guns would often be placed on heavy turrets in the future, the experiment of double-deck heavy and semi-heavy guns was not to be repeated. There had been 'a long, scientific and sometimes acrimonious discussion' (according to a contemporary newspaper) debate in Navy circles about the placing, as well as of the respective merits and deficiencies of 305mm

A stern view of the *Kearsarge*, showing the original aftermast, with its open fighting tops and flags flying. A pair of searchlights is mounted beneath the lower platform.

Draught

American naval tradition fondly remembered the heavily armed frigate, packing a big punch within a relatively small hull, but more up-to-date considerations weighed with the naval planners. Relations with Mexico and other American states were of more concern than global strategy in 1900, and it was to make operations in shallow waters off Mexico possible that *Kearsarge*'s draught was limited to 7.16m (23ft 6in); though fully loaded, including coal to its maximum capacity of 1442 tonnes (1590 tons), the ship would undoubtedly have exceeded that. The restriction of draught was ultimately lifted, but it contributed to the double-turret design, as an attempt to save overall weight even if it imposed a heavy strain on the hull at these points. It was not until the 18,143 tonne (20,000 tons) *Delaware* class, laid down in 1907 and completed in 1910, that the draught of a US Navy battleship exceeded 7.47m (24ft 6in).

(12in) and 330mm (13in) guns. But the bigger gun had a punch that was estimated by the Bureau of Ordinance as 30 per cent more powerful (the weight of a shell rises at least the cube of the increase in calibre) and in tests the 330mm (13in) gun pierced 356mm (14in) armour at 1372m (1500yd) while the 305mm (12in) shells failed to break through. To many foreign observers, American battleships continued to be over-gunned for their size. *Kearsarge* presented a rather austere appearance, with the long blank sides of the casemate and two tall thin funnels between pole masts.

Years of activity

Kearsarge (incidentally the only US battleship not to be named after a State of the Union) was deployed as flagship of the North Atlantic Squadron. In the summer of 1903 it served as flagship of a special squadron on a goodwill visit to European countries, including Great Britain and Germany. On 1 December it left New York for Guantanamo, Cuba, for the formal handover of the US base there. Further goodwill visits were to Spain and Greece in June–July 1904. USS *Maine* took over as Atlantic flagship on 31 March 1905 but *Kearsarge* remained with the fleet until joining the 'Great White Fleet' which showed the American flag (and America's new naval strength) around the world between 16 December 1907 and 22 February 1909.

In September 1909 a long modernisation process was begun at the Philadelphia Navy Yard and *Kearsarge* was not back in commission until 23 June 1915, again with the Atlantic Fleet, landing marines at Veracruz, Mexico, during an operation between 28 September 1915 and 5 January 1916. On reserve from 4 February until April 1917, it then served as a training ship while also patrolling the east coast between Massachusetts and Florida.

New role

On 10 May 1920 *Kearsarge* was decommissioned and work began on converting it into a crane ship, its superstructure replaced by a giant 226-tonne (250 tons) revolving crane for dockyard work and salvage. 'Blisters' were built on to the hull to improve stability when lifting. In this capacity it raised the sunken submarine *Squalus* which had foundered off the New Hampshire coast on 23 May 1939. The name was transferred on 6 November 1941 to a new aircraft carrier, but as Crane Ship No. 1 it continued to serve after 1945, first at San Francisco, then at Boston, where it was finally struck on 22 June 1955, and sold for scrap on 9 August that year.

Mikasa (1902)

Admiral Heihachiro Togo's flagship, and Japan's largest battleship in the Russian-Japanese War of 1905, the British-built *Mikasa* was involved in two fierce battles, in the Yellow Sea and at Tsushima. Later sunk, and raised, it is the only surviving battleship of the pre-*Dreadnought* era.

In 1898 Japan had embarked on a substantial programme of fleet enlargement, despite having limited facilities for building capital ships. Diplomatic tensions with Russia following the war between Japan and China in 1894–95 were a prime reason for the build-up, whose requirements included six battleships. *Mikasa* was ordered from the Vickers yard at Barrow in Furness, England, laid down on 24 January 1899, launched on 8 November 1900, and

Mikasa was the second of three ships in the *Asahi* class, all built in British yards between 1900 and 1902. In general they were very similar to the British Majestic class 'pre-Dreadnoughts' in design.

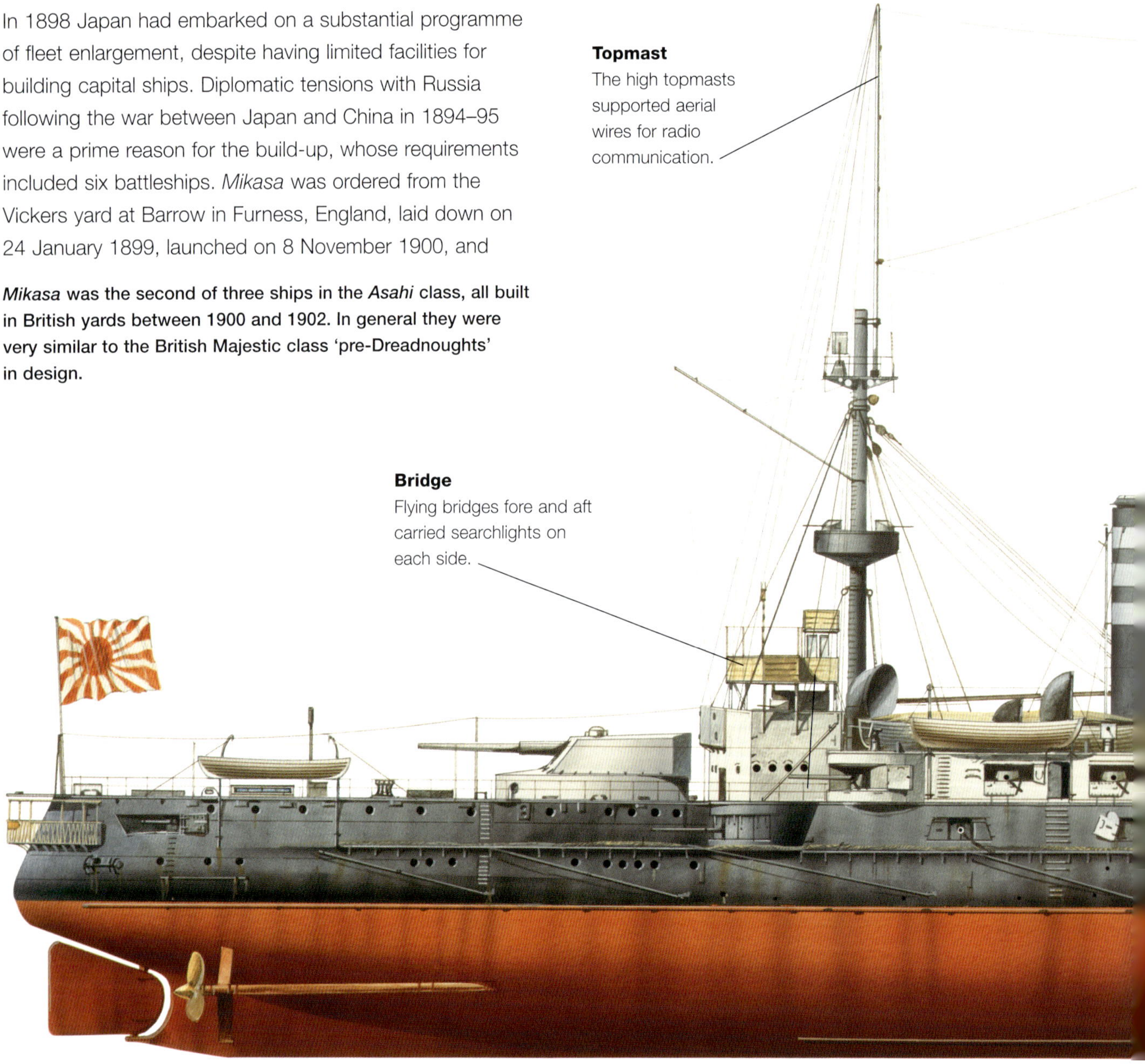

Paint colour
Mikasa as preserved is painted in 'battleship grey' but this paint finish, with the triple-stripe funnels, was ordered for its original completion.
Emblem
The emblem on the bow is the Japanese imperial symbol of the chrysanthemum.

Radio communication at Tsushima

Radio very quickly became a vitally important part of a capital ship's operating system. Early ship-to-ship radio could transmit messages by the Morse code over 80km (50 miles) but at first only on a one-to-one basis. Thus the message, 'Enemy ship in sight' was relayed from the auxiliary cruiser *Shinano Maru* to the cruiser *Itsukushima* and then to *Mikasa's* wireless-telegraphy room in the aft superstructure. Admiral Togo could know the Russian ships' position and direction with a precision that was previously impossible at other than close range. Tsushima was the first naval battle in which wireless communication (in use on both sides, the Japanese using their own system, the Russians using German *Telefunken* equipment) played a significant part. The British Royal Navy introduced wireless communication from 1900. By 1907 the American 'Great White Fleet' was equipped with radio-telephones, but this was ineffective and abandoned until a new system was introduced in 1916.

commissioned into the Imperial Navy on 1 March 1902 as its largest and most potent battleship. Total cost was £880,000.

Design and specification

The design was an enlarged version of the Royal Navy's *Majestic* class, whose plans were drawn up in 1897–98. A slightly raised forecastle topped a ram bow, and its four 305mm (12in) gun turrets were mounted in barbettes at each end of a casemate which housed ten 152mm (6in) guns, with an upper level armed with a further four 152mm (6in) and twenty 12-pounder guns, four of them mounted on the 152mm turrets. Eight 3-pounder, four 2.5-pounder and eight Maxim multi-barrelled machine guns completed the batteries, and four submerged 457mm (18in) torpedo tubes were fitted. The 305mm (12in) guns were of Vickers' own design, in typical British-type turrets with sloping faces and roofs through which armoured sighting hoods protruded.

Four Scottish-made Barr & Stroud FA3 coincidence rangefinders, effective up to 7315m (8000yd) were fitted, and the turrets had 24-power magnification gunsights. Krupp cement armour, which had replaced Harveyised nickel steel as the most shell-resistant protection, was fitted to the steel plates, and the internal armoured deck sloped downwards fore and aft to meet the ram and to cover the steering gear. *Mikasa* had two pole masts fitted with derricks and radio aerials.

Russo-Japanese War

Mikasa's aggressive and defensive abilities were to be heavily tested in the Russo-Japanese war of 1904–05: a campaign watched with intense interest by the other naval powers because there had been no major naval battle for decades. On 10 August 1904 the Japanese fleet under Admiral Heihachiro Togo, with *Mikasa* as his flagship, engaged the Russian Pacific fleet in the fast-moving Battle of the Yellow Sea. No ships were sunk, but the Russians were driven into the shelter of Port Arthur.

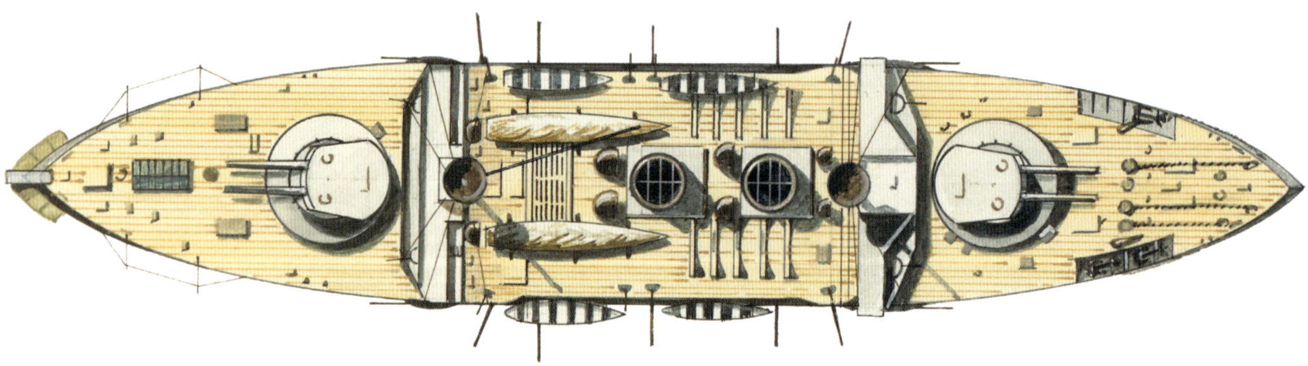

Deck plan of *Mikasa*. The anchor cables come out on the open forecastle deck. A derrick was also mounted on the foremast (starboard side).

Mikasa in drydock at the Royal Navy yard, Portsmouth, England, early 1902, before making the voyage to join the Imperial Japanese Navy.

Specification

Dimensions	Length 131.7m (432ft), Beam 23.2m (75ft 6in), Draught 8.28m (27ft 6in), Displacement 13,789 tonnes (15,200 tons)
Propulsion	25 Belleville watertube boilers, 2 vertical triple-expansion engines, 11,185kW (15,000hp), 2 screws
Armament	4 305mm (12in) guns, 14 152mm (6in) guns, 20 12-pounder, 8 3-pounder, 4 2.5-pounder guns, 4 457mm (18in) torpedo tubes
Armour	Belt 229–102mm (9–4in), Bulkheads 305mm (12in), Deck 76mm (3in), Barbettes 356–254mm (14–10in), Main turrets 254–203mm (10–8in), Lower deck redoubt and battery 152mm (6in)
Range	13,000km (7000nm) at 10 knots
Speed	18 knots
Complement	830

Mikasa sustained around 20 shell strikes, the aft main turret was put out of action and it lost its radio links, but its machinery was undamaged.

Nine months later, when the Russians had despatched their Baltic fleet 33,000km (18,000nm) to the Pacific, the major battle of Tsushima was fought in the waters between Japan and Korea on 27–28 May 1905, with *Mikasa* again as Togo's flagship. While the Japanese had only four battleships against the Russians' eight, they had a much larger number of cruisers, destroyers and torpedo boats – altogether 89 vessels against the Russians' 28. The result was a Nelsonian-scale victory for Togo, with seven Russian battleships and fourteen other enemy vessels sunk, for the loss of three torpedo boats. Though *Mikasa* took some 40 strikes, including 10 from 305mm (12in) shells, no major damage was done.

Tsushima was the first battle in which radio communication between ships was important. It also demonstrated the value of heavy guns, long-range fire (with range-finders to guide the guns), and speed. A battleship's secondary armament of 203mm (8in) or 152mm (6in) guns was shown to be ineffective and perhaps even detrimental to successful combat. It was the 305mm (12in) guns, firing at relatively long range, which decided the issue.

Explosion and sinking

Only a few weeks later, anchored at the Sasebo base, *Mikasa* suffered a fire and internal explosion, and sank in relatively shallow water on 11 September, killing 339 of the crew. Refloating was achieved on 8 August 1906 and the ship was repaired at Maizuru Naval Arsenal and restored to active service in 1908, with new 305mm (12in), 45 calibre guns that fired shells faster and farther. It remained in active service, though progressively reduced in status until by 1 September 1921 it was rated as a coastal defence ship, first-class. Shortly after that, it grounded off the Russian coast while supporting Japanese forces intervening in Russia's post-revolutionary civil war. Though repaired, that marked the end of *Mikasa*'s active service. It was decommissioned late in 1921. Since 1925 it has been a museum display ship at Yokosuka (restored 1958–61).

Tsessarevich (1903)

***Tsessarevich* ('Crown Prince') was built in France and closely resembled French battleships of the time, with outwards-bulging sides and large armoured fighting tops in the masts. It survived the end of the Tsarist era to receive the name 'Citizen' under the Soviet regime.**

With three widely separated coastlines to protect, Russia required a large naval force. The majority of its battleships were built in Russian yards, but two of the ships that fought the Japanese were foreign-built, *Retvisan* at Philadelphia, and *Tsessarevich* at La Seyne-sur-Mer, France.

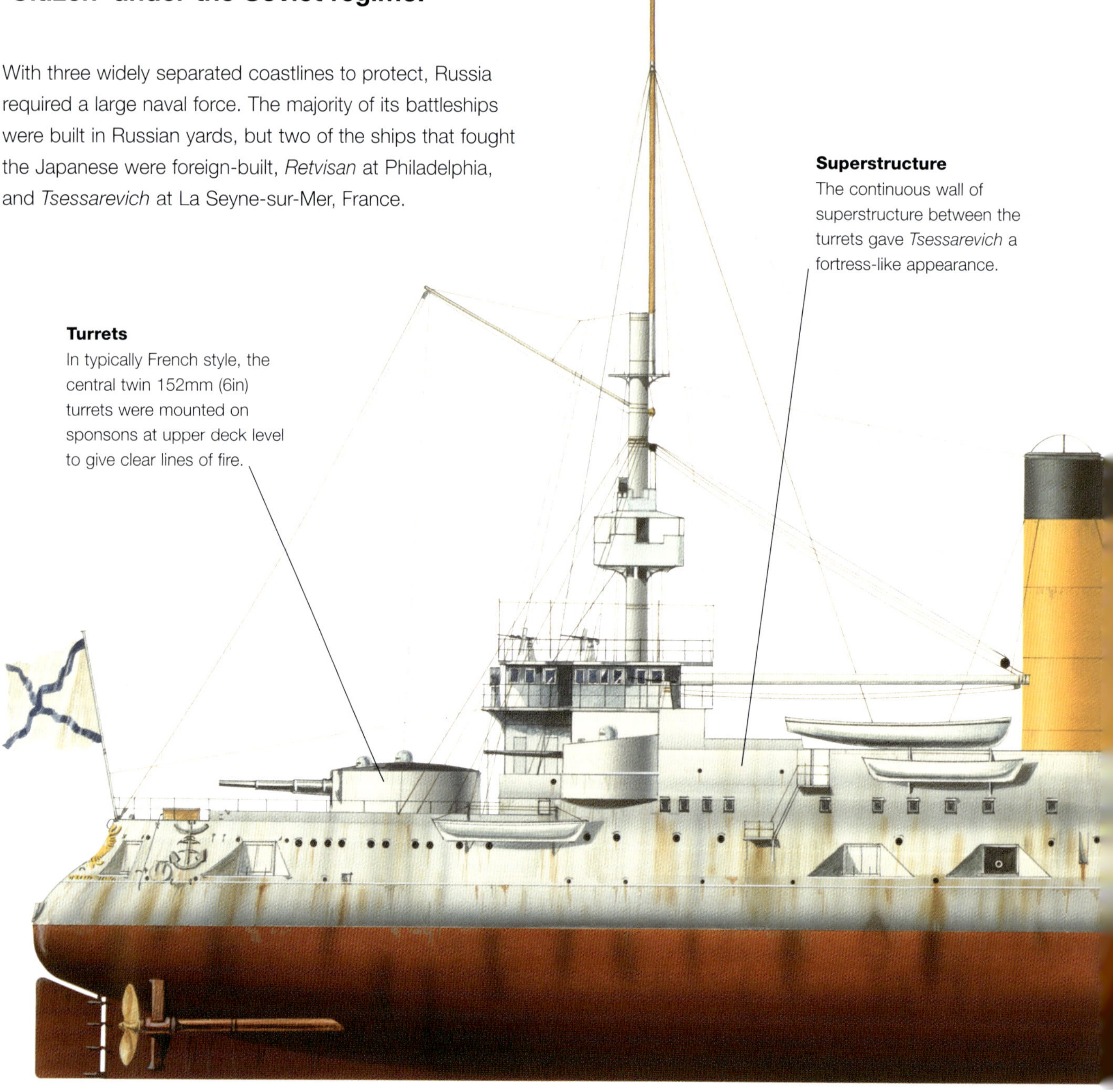

Superstructure
The continuous wall of superstructure between the turrets gave *Tsessarevich* a fortress-like appearance.

Turrets
In typically French style, the central twin 152mm (6in) turrets were mounted on sponsons at upper deck level to give clear lines of fire.

Tsessarevich did not take part in the Battle of Tsushima (1905) and underwent numerous modifications following the disastrous Russian defeat.

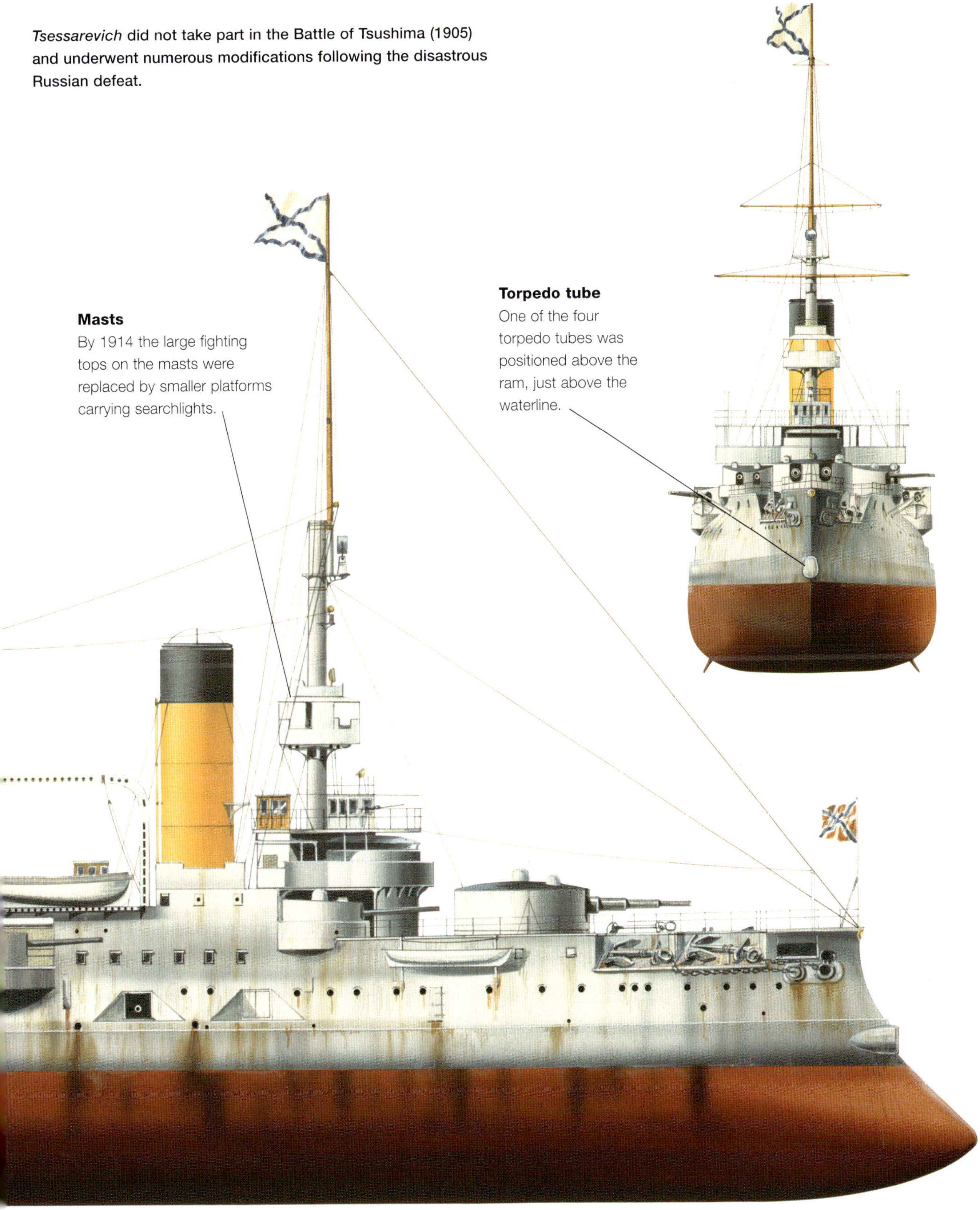

Specification

Dimensions	Length 118.5m (388ft 9in), Beam 23.2m (76ft 1in), Draught 8.5m (27ft 11in), Displacement 13,122 tonnes (12,915 tons)
Propulsion	20 Belleville boilers, 2 vertical triple-expansion engines, 12,155kW (16,300hp), 2 screws
Armament	4 305mm (12in) guns, 12 152mm (6in) guns, 16 75mm (3in) guns, 4 47mm (1.9in) 3-pounder guns; 6 457mm (18in) torpedo tubes
Armour	Belt 230–150mm (9–5.9in), Main turrets 254mm (10in), Secondary turrets 150mm (5.9in), Conning tower 254mm (10in), Deck 57mm (2.25in)
Range	10,186km (5500nm) at 10 knots
Speed	18 knots
Complement	779

The immensely tall funnels, broad yards and bulging sides of *Tsessarevich* made a strong impression on its visit to Portsmouth, England, in 1913.

Laid down on 7 August 1899, launched on 23 February 1901, it was commissioned on 3 September 1903. Though the Russian Marine Technical Committee had produced a specification, the design was heavily influenced by French practice. The same yard had built *Jauréguiberry* and the design chief, Lagane, produced a ship on very similar lines, though larger, with a bulging tumblehome hull and a narrow superstructure enabling sponson-mounted side guns to fire fore and aft. Greater dimensions enabled the main and secondary guns to be mounted in twin turrets.

The hull had Lagane's longitudinal bulkhead, dividing the armoured deck, and intended for anti-torpedo protection. It separated the two engine compartments, making it possible for one to continue functioning if the other should be put out of action. But it also meant that inflowing water could cause the ship to list or even capsize. Nevertheless, this bulkhead, installed for the first time in *Tsessarevich*, was incorporated in many post-*Dreadnought* battleships. Krupp-type armour was fitted, the reduced thickness compared with earlier ships demonstrating its resistance power. The guns were manufactured by Schneiders of Le Creusot to the design of Canet, the leading gun designer of the time. Two pole masts were fitted, the aftermast having a derrick attached. Gantry-type davits between the funnels launched the ship's boats.

Deployment

Tsessarevich had been ordered for the Pacific fleet and was immediately deployed to Port Arthur as flagship. In the unannounced attack on Port Arthur (8 February 1904) which preceded Japan's declaration of war against Russia, it was struck by a torpedo and remained out of action for several weeks. On 10 August it led the fleet out to fight the Japanese, who were attempting to blockade the port, and the day-long battle of the Yellow Sea began, with both fleets firing from extreme range of around 9100m (10,000yd). At 18:00 a shell splinter killed Admiral Vitgeft on the *Tsessarevich*'s bridge and a few minutes later further hits virtually wrecked the bridge and jammed the steering wheel, sending the ship into a sharp turn. Mistaking this for a planned movement, other ships followed the crippled flagship, breaking the Russian line and creating confusion among the captains. The advent of darkness and depleted ammunition stocks forced the Japanese Admiral Togo to break off the engagement and the Russian fleet was able to return to Port Arthur, except for *Tsessarevich* which, escorted by three destroyers, was able to make the Chinese port of Tsingtao (Qingdao), then an enclave of German control (from 1898 to 1914), where it was interned until the war ended. Thus *Tsessarevich* did not participate in the Battle of Tsushima.

One of the results of that battle was that the tumblehome style of construction – used also in the five *Borodino*-class battleships, built in Russia, which had been partially modelled on *Tsessarevich* – was abandoned. Three out of the four participating *Borodino*-class ships were sunk at Tsushima and the design was blamed for instability in action, especially when water entered the hull. Tumblehome was dropped from the naval architects' resource-list until the twenty-first century and the US *Zumwalt* destroyer.

1905 and beyond

In 1905 *Tsessarevich* was transferred to the Baltic Fleet, which had lost its battleships at Tsushima, and it continued to be based at Kronstad. A courtesy visit was made to Portsmouth, England, in 1913. In World War I it was engaged on Baltic patrols. With the end of the Tsarist regime in April 1917, the ship was renamed *Grazhdanin*, 'Citizen', and a sailors' committee replaced the former command structure. In 1917 it was with the Gulf of Riga Squadron along with the *Borodino*-class battleship *Slava* (commissioned in October 1905) and a force of cruisers and destroyers.

On 17 October 1917, the Riga Squadron was attacked in Moon Sound, between the island of Muhu and the Estonian coast, by a German force including the *Dreadnought*-type battleships *König* and *Kronprinz Wilhelm*. *Slava*, heavily damaged, was scuttled by its crew, but *Grazhdanin* survived. By 1918 it was obsolete and probably unworkable. Hulked in that year, it was sold to Germany for scrapping in 1924.

Gun calibres

The 305mm (12in) gun was the standard big gun of the Imperial Russian Navy from 1886 until 1906. Originally modelled on a German Krupp gun of 35 calibre, from 1892 an improved version was used, based on French Canet designs supplied to the Russian Obukhov Works at St Petersburg by Schneiders. These were of 40 calibre. Calibre is used in two senses, first as an expression of the diameter of the barrel (the bore) and projectile, as in a 305mm (12in) gun firing a shell of the same diameter; second as an expression of the barrel's length as a multiple of the bore. A 305mm (12in) 40 calibre gun has a barrel length from breech-face to muzzle of 40 times the bore: 12.2m (40ft). The 40 calibre guns fitted to *Tsessarevich* had greater velocity, range and precision than the older 35 calibre guns: 10.67m (35ft) firing shells of the same weight.

Deck plan: the laterally-mounted 75mm (3in) guns were situated so close to the waterline as to be of little offensive value.

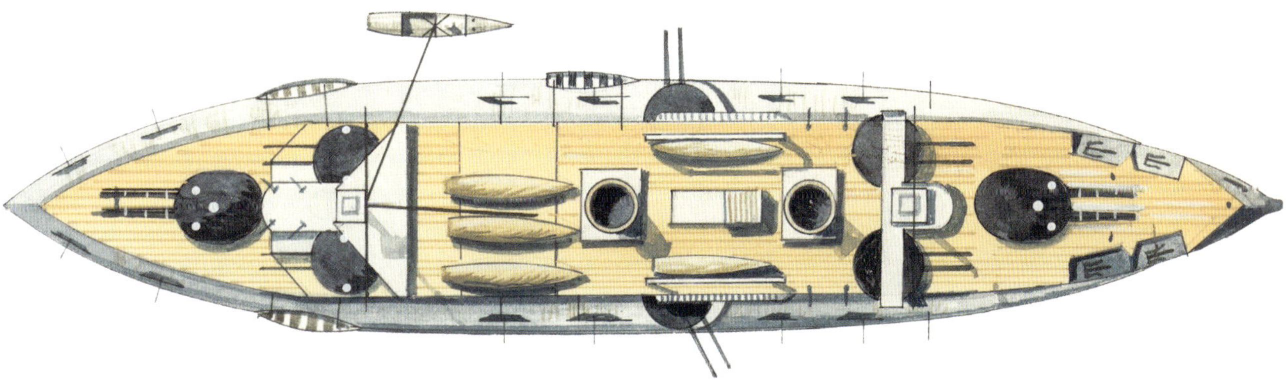

Dreadnoughts and Battleships

Four decades of the twentieth century saw the capital ship reach a peak in size, armament, protective armour, speed – and cost. Their immense destructive power was very real, and helped make them potent symbols of national prestige and pride. But this was the final era – by the 1940s, it was clear that battleships no longer ruled the waves.

Opposite: 'Gobs and Guns'. A scene aboard the USS *Texas*, just back from foreign waters circa 1918, showing the 'gobs' enjoying a little fun on the 356mm (14in) main guns.

Dreadnought (1906)

A new era of fast, big-gun battleships was inaugurated by the appearance of HMS *Dreadnought.* Similar ships were already being planned or considered by other navies and the effect, instead of assuring British dominance, was to start a naval arms race.

When the Battle of Tsushima was fought on 27 May 1905, the British Navy was already contemplating a battleship in which many lessons taught by the battle seemed to have been anticipated. *Dreadnought* was laid down at Portsmouth on 2 October 1905, launched on 10 February 1906, and completed in December 1906. The cost was £1,783,883. Speed of construction was remarkable, made possible by pre-assembly of the materials, but several years of debate went before it.

Genesis of *Dreadnought*

In 1903 the Italian engineer Vittorio Cuniberti had suggested that the 'ideal battleship' for the Royal Navy was a 15,422 tonne (17,000 tons) ship with 12 305mm (12in) guns, 305mm (12in) armour, and capable of 24 knots. Admiral Lord Fisher had already initiated a project for what he called 'HMS Untakeable', whose prime feature was 'the most powerful and powerfully arranged armament', with a view to long-range fire rather than the

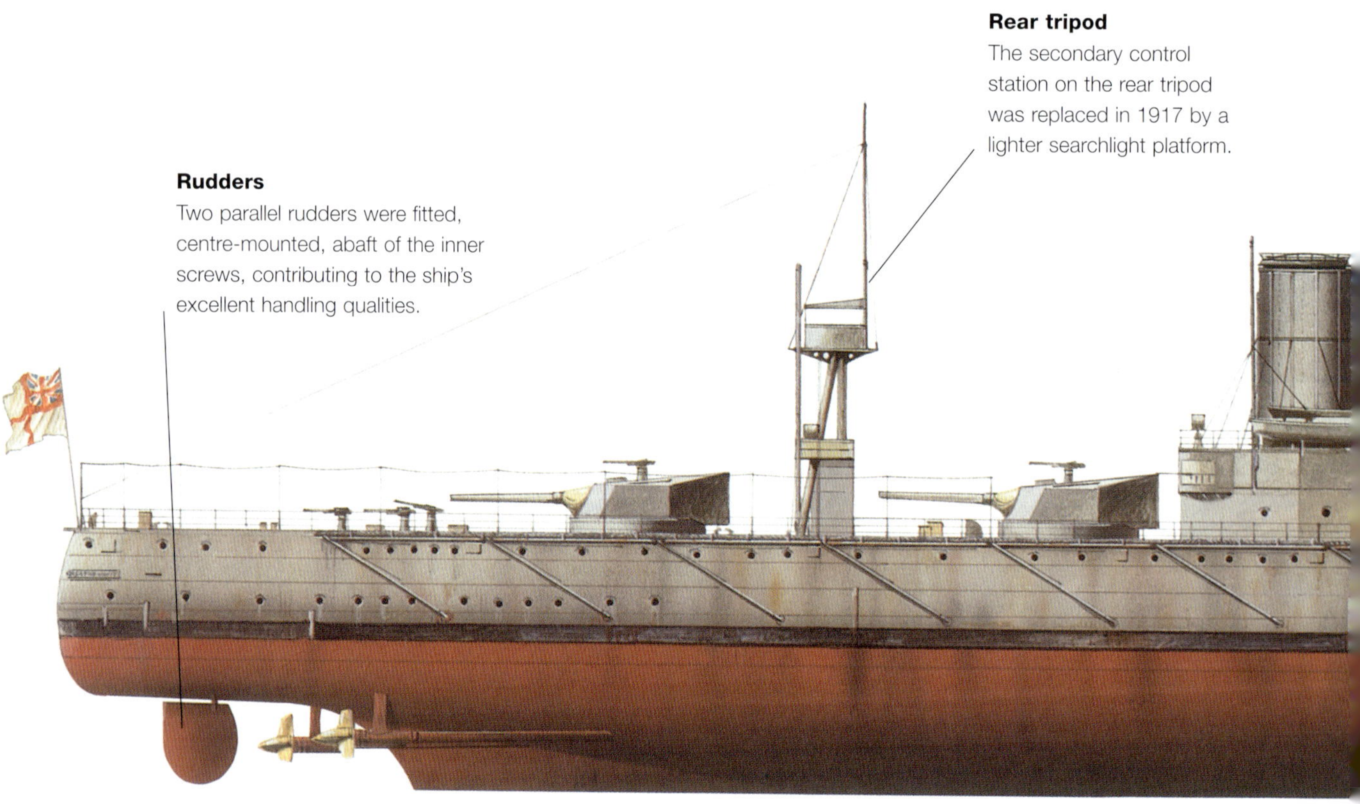

Steam turbines, quadruple screws, and good armour protection, combined with 10 305mm (12in) guns, made *Dreadnought* a world-beater.

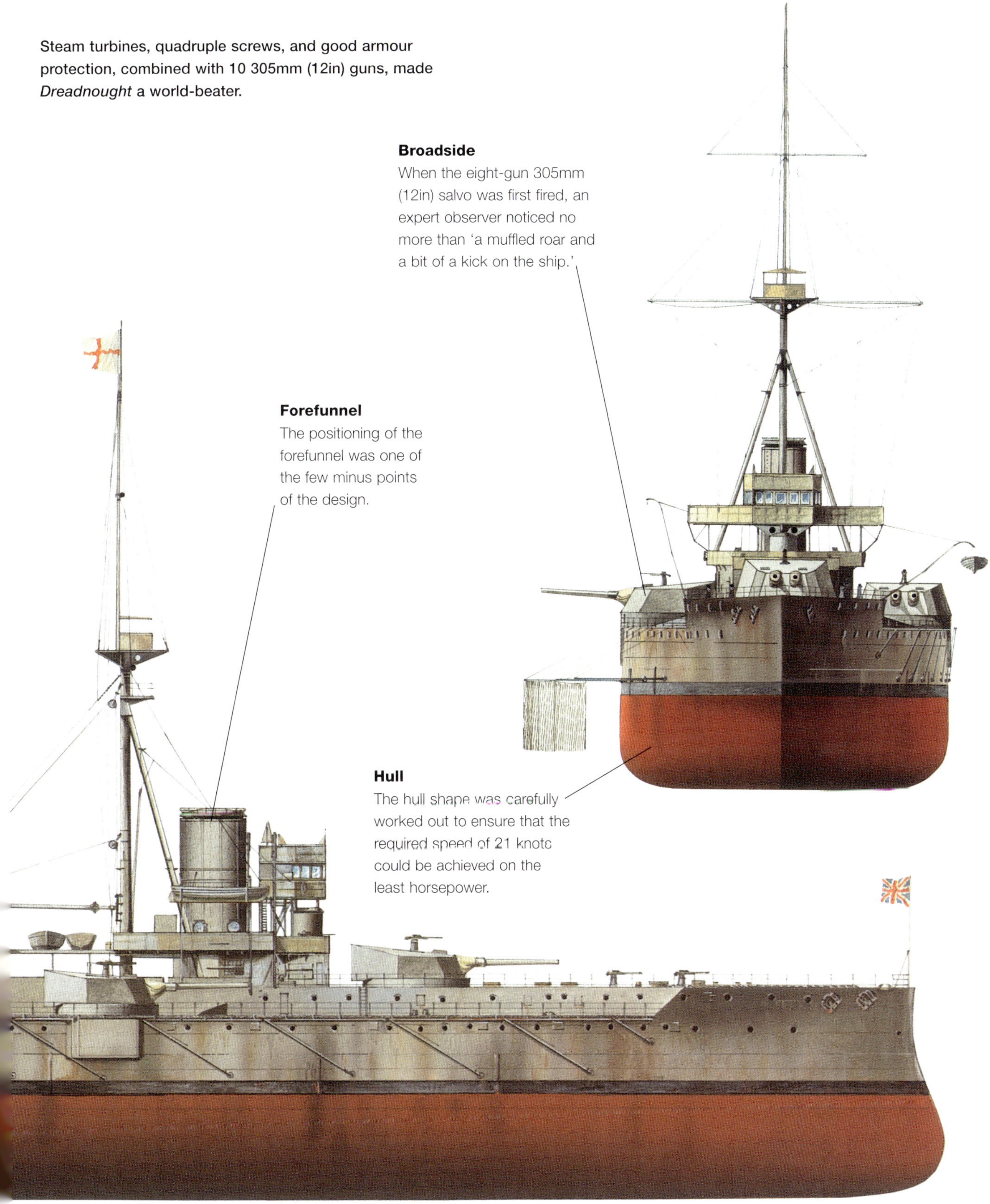

Broadside
When the eight-gun 305mm (12in) salvo was first fired, an expert observer noticed no more than 'a muffled roar and a bit of a kick on the ship.'

Forefunnel
The positioning of the forefunnel was one of the few minus points of the design.

Hull
The hull shape was carefully worked out to ensure that the required speed of 21 knots could be achieved on the least horsepower.

Specification

Dimensions	Length 160.4m (527ft), Beam 25m (82ft), Draught 8.1m (26ft 6in), Displacement 16,238 tonnes (17,900 tons). Full load 19,817 tonnes (21,845 tons)
Propulsion	18 Babcock & Wilcox boilers, 4 Parsons turbines, 19,649kW (26,350hp), 4 screws
Armament	10 305mm (12in) guns, 27 12-pounder guns, 5 457mm (18in) torpedo tubes (submerged)
Armour	Belt 279–102mm (11–4in), Bulkhead 203mm (8in), Barbettes 279–102mm (11–4in), Turrets 279mm (11in), Conning tower 279–203mm (11–8in)
Range	12,260km (6620nm) at 10 knots
Speed	21.6 knots
Complement	773

1000–3000m/yd considered normal at the time. Fisher was appointed First Sea Lord in October 1904 and convened an expert committee which proceeded to draw up six alternative preliminary designs. As built, based essentially on 'Design H', there was a long flush main deck of 5.8m (19ft) freeboard, with a forecastle deck raised to 8.5m (28ft). Ten 305mm (12in) guns in double turrets were disposed in symmetrical array, three on the centreline and one each to port and starboard just abaft of the main deckhouse and leading funnel. This funnel, unusually and unsuitably, was place ahead of the foremast, making the fighting top filthy with smoke and soot. Another novelty, not particularly appreciated, was the berthing of officers in the forward part (close to their work posts) and of the men aft.

Exactly 4536 tonnes (5000 tons) of armour was applied, and armour plating was extended, at diminishing thickness, along the entire sides. Within the hull a system of fully watertight compartments was rigorously applied, with no doors or openings and each compartment having independent ventilation, pumps (electrically driven) and drainage. Lifts were fitted for giving quick access between the machinery compartments. The aim was that the ship could safely sustain two torpedo strikes. Magazines were located centrally, well away from the sides.

In order to speed up construction, the 305mm (12in) guns were 'lifted' from two other ships then building, *Lord Nelson* and *Agamemnon*, which were not launched

Gunnery

Led by a few keen senior officers, the Royal Navy took great steps forward in improving its gunnery from the late 1890s. *Dreadnought*'s capacity for firing heavy salvoes demanded a high degree of fire control to make it effective. For effective fire control, correctly establishing the target's distance was essential, and crucial items were optical rangefinders, developed from the 1890s by the Barr & Stroud company. Other new instruments were devised, including range projectors (Dumaresqs, an early form of computer which related the movement of the firing craft and target) and range clocks, calibrated up to 11,000m (12,000yd). The German optical firm of Zeiss was also producing a range of similar instruments. A central director could be aimed at the target and the required elevation and training angle automatically calculated. Plotting

boards, with scales and protractors, were brought into use. Even so, a 10 per cent hit rate in a long-range naval battle was good; 90 per cent of ammunition vanished into the sea.

Fuel: 1016 tonnes (1120 tons) of oil fuel was carried in the bottom under the machinery spaces. A maximum 2631 tonnes (2900 tons) of coal could be carried in the bunkers.

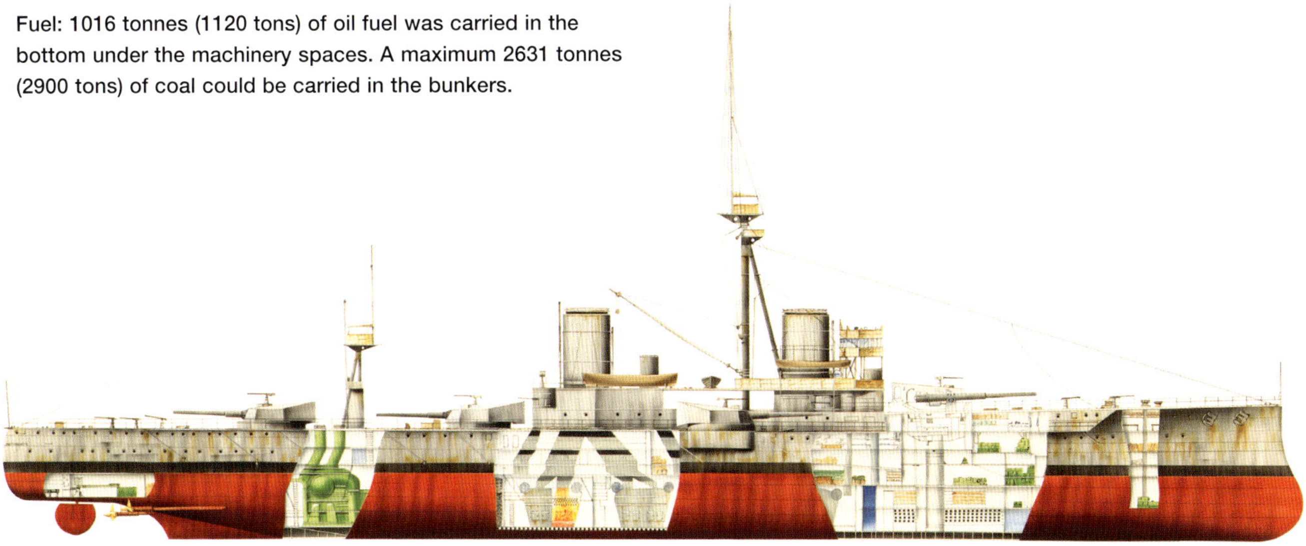

until after *Dreadnought*, effectively becoming £3,000,000 worth of instantly-obsolescent sea-power.

Fastest battleship

The most revolutionary aspect of the new ship was its motive power. Turbines had never been applied to a battleship before, though they were in use on ocean liners and fast packet boats. The decision proved a triumphant success. *Dreadnought* was faster than any other battleship, mechanically more reliable, more economical on fuel, and provided infinitely better working conditions in the engine-room. In addition the engine space was more compact, sat lower in the hull, and gave substantially more power per tonne of weight. Total weight of machinery was 1079 tonnes (1190 tons). Turbines could not be reversed, but a stern turbine was placed on each of the four shafts, one high- and one low-pressure on each side, and *Dreadnought* was at least as easy to handle as any other battleship and had very good manoeuvring qualities, assisted by twin rudders placed immediately abaft the inner propellers.

Dreadnought was assigned to the Home Fleet, and from January–March 1907 went on a shake-down cruise to the Mediterranean, then across to Trinidad. Its return journey of 11,265km (7000 miles) was accomplished with no mechanical hold-ups and at an average speed of 17.5 knots. No warship before had been capable of this. From April 1907 to May 1912 it was flagship of the C-in-C. Home Fleet, then flagship of the 4th Battleship Squadron until May 1916. In World War I it sank *U-29* by ramming in the North Sea on 18 February 1915: an ironic exploit considering that a ram bow had been deliberately omitted from its design as something not required in a modern battleship. From July 1916 it was flagship of the 3rd Battleship Squadron at Sheerness, but briefly rejoined the Grand Fleet between March and August 1918. In February 1919 it was placed on reserve at Rosyth, and stricken on 31 March 1920. In May 1921 it was sold for scrapping, at Inverkeithing.

Deck plan: an eight-gun broadside salvo could be fired on either beam. *Dreadnought* was also the first battleship fitted with tripod masts.

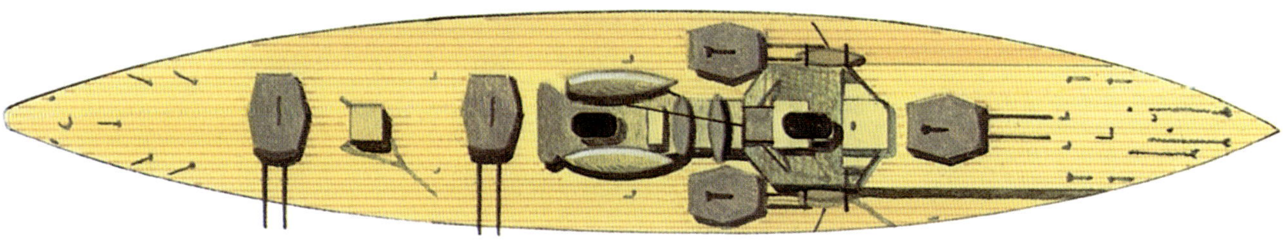

Nassau (1909)

One of the first German big-gun ships to be built after *Dreadnought*, *Nassau* carried a substantial secondary armament as well as 12 main guns. Powered by triple-expansion engines rather than by turbines, it was in action with the High Seas Fleet at Jutland.

Plans for ships of this class had been worked on from March 1904 and the final design was completed in 1906. Four battleships formed the class, with *Rheinland* as the first to be laid down but *Nassau* first to be completed on 1 October 1909, having been laid down at Wilhelmshaven on 22 July 1907 and launched on 7 March 1908. The others were *Posen* and *Westfalen*. They cost around 37.5 million Goldmarks each, and all were in service by May 1910.

Secondary armament

These were large battleships, mounting 12 heavy guns, but unlike *Dreadnought* a large-scale secondary armament was also included, with 12 150mm (5.9in) guns mounted in casemates at a level below the port and starboard main turrets, and 16 86mm (3.4in) guns in side-mounted sponsons on the hull and superstructure. Foremast and fore funnel were very close to each other, and to the deckhouse, with navigation bridge and chart house. *Nassau*'s masts

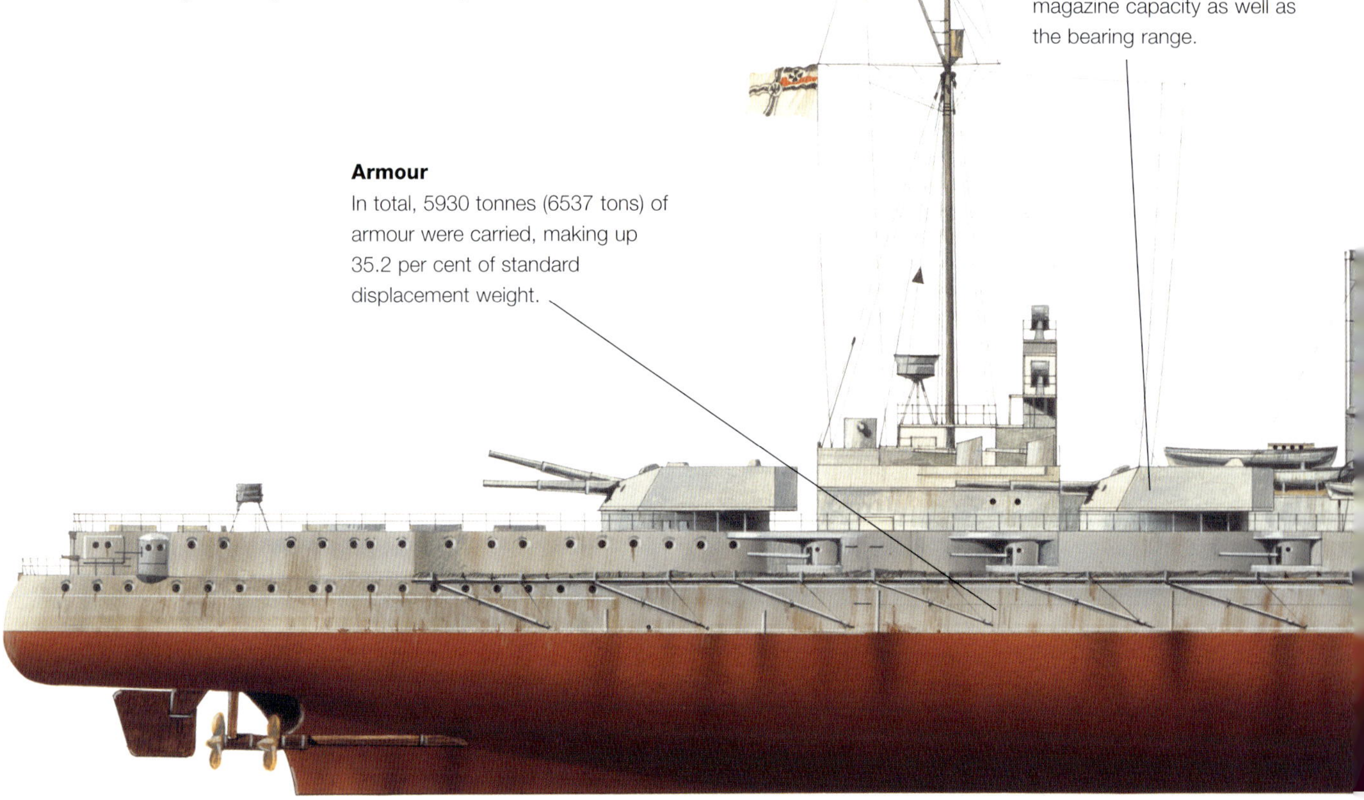

Nassau was a beamy ship (beam 0.18 per cent of length compared to *Dreadnought*'s 0.15 per cent), most of the difference being used for additional protection.

Torpedo nets
Torpedo nets, made of steel mesh, were a fixture of all capital ships from the 1890s up to 1916. Properly disposed, the net hung from waterline level. Inevitably torpedoes were redesigned with 'netcutter' noses and nets in turn were made of heavier-grade mesh.

Boilers
Supplementary oil burners were fitted to the 12 boilers from the end of 1915.

Compartments
Nassau had 16 internal watertight compartments; others in the class had 19.

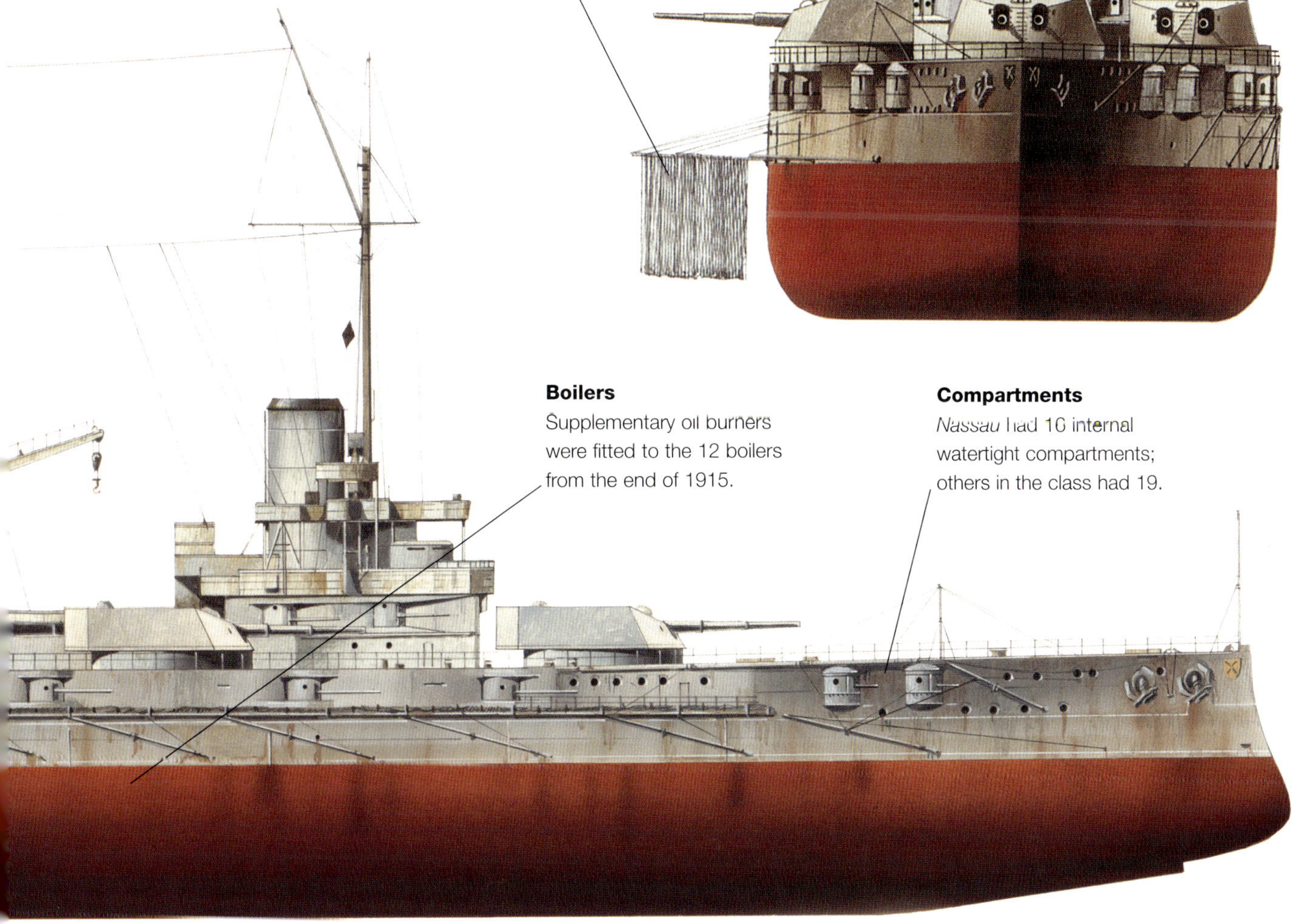

Specification

Dimensions	Length 146.1m (479ft 4in), Beam 26.9m (88ft 4in), Draught 8.5m (27ft 11in), Displacement 17,146 tonnes (18,900 tons)
Propulsion	3 vertical triple expansion engines developing 16,405kW (22,000hp), 3 screws
Armament	12 280mm (11in) guns, 12 150mm (5.9in) guns, 16 88mm (3.5in) guns, 6 450mm (17.7in) torpedo tubes
Armour	Bulkhead 200mm (7.8in), Belt 300mm (11.8in), Tower (forward) 400mm (15.6in), aft 200mm (7.8in), Barbettes and turrets 280mm (11in)
Range	17,408km (9400nm) at 10 knots
Speed	19.5 knots
Complement	963

This contemporary cigarette-card view shows the heavy stern protection below the two original 86mm (3.4in) stern-chaser guns.

had high wireless aerial gaffs set at an angle from the mizzen top of both masts; these were removed in 1911, and during World War I a spotting top was fitted on the foremast. Gooseneck cranes at each side of the aft funnel swung out the boats housed amidships.

The main guns, 280mm (11in), were of smaller calibre than the 305mm (12in) guns being established as the British standard, but extensive testing had convinced the German navy that they were not significantly less effective. They had a barrel calibre of 45 and weight of 47.7 tonnes (52.6 tons) and fired a 305kg (672lb) shell 18,900m (20,669yd) with a 20 degree elevation.

Its best firing rate was three rounds in two minutes. Comparative figures for *Dreadnought*'s guns were: barrel length identical, barrel weight 51.7 tonnes (57 tons), shell weight 385kg (849lb), range 19,000m (20,779yd) at 13 degrees of elevation and a rate of fire of two rounds a minute. The advantage would seem to be with the British, but the German admirals believed in the armour-piercing qualities of their shells.

Nassau and its sister ships had triple expansion engines with water-tube boilers; the first German heavy ship to have Parsons turbines was the battlecruiser *Von der Tann* of

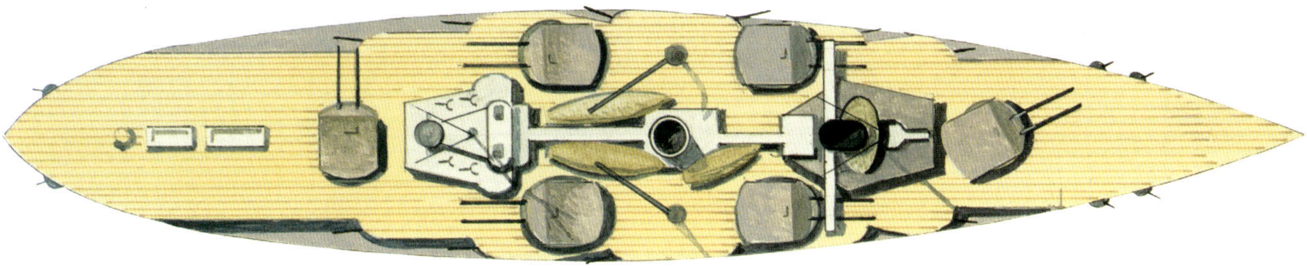

Twelve 280mm (11in) guns were carried, compared to *Dreadnought*'s ten 305mm (12in) guns. Magazine layout beneath the flank turrets was cramped.

1907. Consequently the three boiler rooms and the engine room occupied most of the hull between the masts. In 1915 the boilers were adapted to burn an oil-coal mix, with the oil sprayed above the burning coal. Oil tanks with a capacity of 142 tonnes (157 tons) were installed. The German designers set great store by good underwater protection and *Nassau*'s hull had 16 watertight divisions, with the placing of armour on the class done on a scientific basis. But the underwater lines had to be modified after sea experience. It had been supposed that the wide beam and the lateral placing of heavy guns would make a stable ship, but in some North Sea swells they rolled violently and bilge keels had to be fitted.

In August 1914 *Nassau* was one of the eight ships of Battle Squadron I of the High Seas Fleet (there were three squadrons with a total of 26 battleships). Wartime modifications apart from those already noted included the removal of the stern-mounted 86mm (3.4in) guns in 1915 and the removal of all the others in 1916 to be replaced by four AA guns of the same calibre.

Battle of Jutland

The ship saw no action until an unsuccessful sortie into the Hoofden (North Sea off the Dutch coast) on 15–16 March 1916. On 24 April 1916 it escorted a squadron of battlecruisers to bombard the English coastal towns of Lowestoft and Yarmouth. In the Battle of Jutland, 31 May, it was hit twice by shellfire and ran against the British destroyer *Spitfire* in an attempt to sink it by ramming, but all damage was repaired by 10 July. Subsequently *Nassau* made three further sorties into the North Sea without any positive result; on the last occasion, with other ships of the Squadron including *Westfalen* (pictured below) and *Posen* reaching the latitude of Stavanger (23 April 1918). Not among the ships scuttled at Scapa Flow, it was stricken on 5 November 1919. Intended to go to Japan as war reparation, it was sold by the Japanese government to a British company who had it scrapped at Dordrecht in the Netherlands in June 1920.

South Carolina (1910)

Striking power and staying power were seen as more important than speed by the USN. This two-strong class mounted a formidable eight-gun 305mm (12in) broadside, and their superfiring turrets were to become the standard arrangement for capital ships.

Though *Michigan* was commissioned two months earlier, *South Carolina* had already been allocated the pennant number BB25, and its sister ship was BB26. Ordered as first-class battleships, their standard displacement was restricted to 14,515 tonnes (16,000 tons) by Congressional mandate, though fully loaded they were closer to 16,329 tonnes (18,000 tons).

Design ingenuity

The weight limitation brought out the maximum ingenuity of the American designers as they worked out how to attain maximum firepower with a broadside of eight big guns. Their details had been fixed before *Dreadnought*'s, but they were not completed until well after the British ship. *South Carolina* was laid down at Cramp's yard, Philadelphia, on 18 December 1906, launched on 11 July 1908 and commissioned into the Navy on 1 March 1910. With them, what became the classic gun arrangement of the battleship

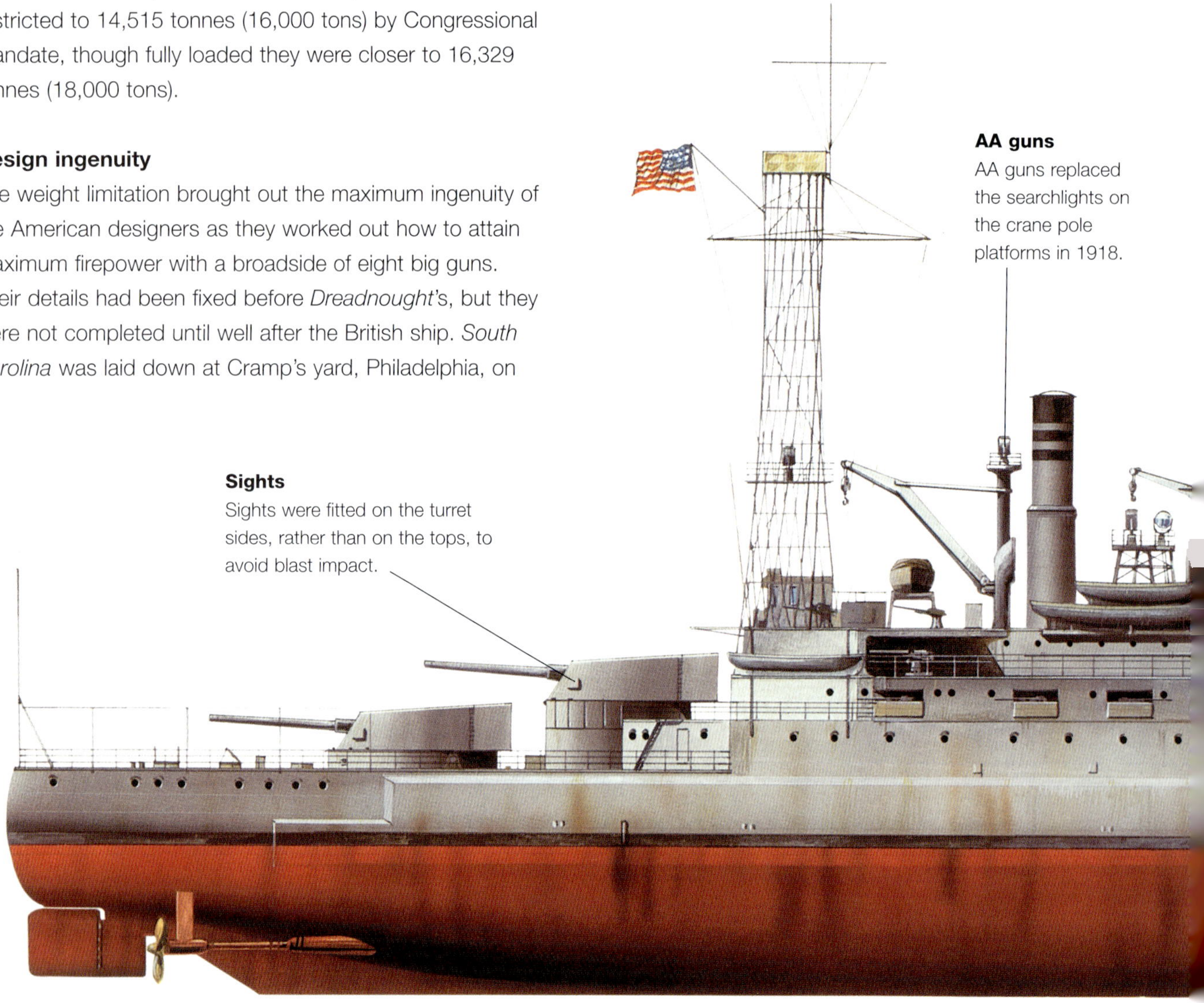

AA guns
AA guns replaced the searchlights on the crane pole platforms in 1918.

Sights
Sights were fitted on the turret sides, rather than on the tops, to avoid blast impact.

The gun layout did away with the need for wing turrets, making better use of magazine space and improving magazine protection, and still enabling an eight-gun broadside to be fired.

Side armour

Side armour reached a maximum thickness of 305mm (12in) alongside the main magazines.

Turret design

Drawn up before HMS *Dreadnought,* this was the first battleship planned with all its big guns in centreline turrets (though the first design incorporated two single turrets to port and starboard).

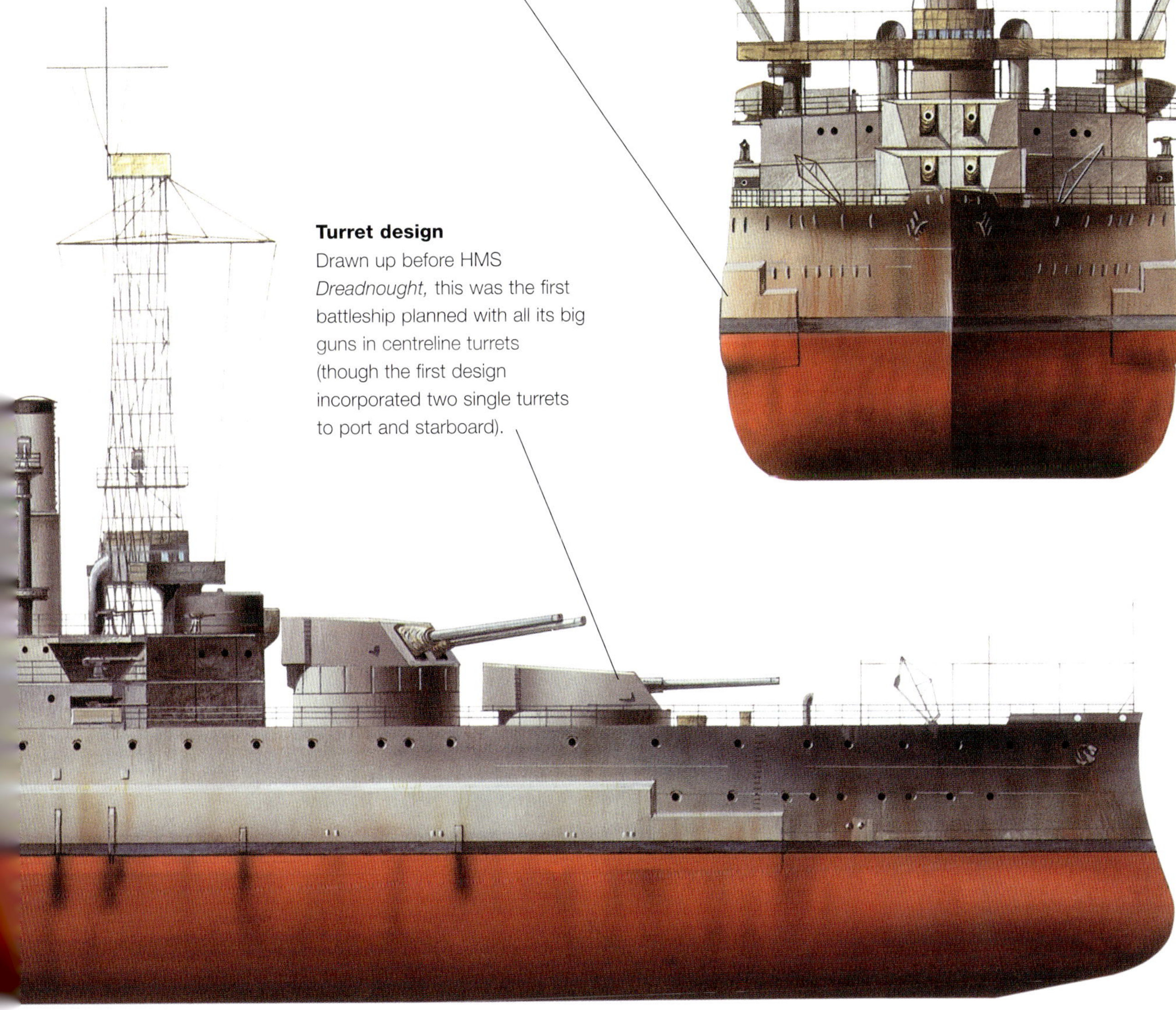

The basket mast

BB25 and 26 were the first US ships to carry the so-called lattice masts, often referred to as basket masts and unfavourably compared to electricity pylons, which were to typify US battleships until the 1930s. In fact they were multitubular steel towers in which the topmost and bottom retaining rings were turned in opposite directions, resulting in a rotary hyperboloid form. It was felt that the nature of the structure would protect fire control positions from the shocks and vibrations that were transmitted through the traditional form of mast.

Tested in gunnery practice, the structure proved stable against shell hits, though *Michigan*'s forward mast was brought down in a gale in 1918. The top platform was exposed and unsuitable as a control post or gun-platform, and the design could not support an armoured fire-control position, so the basket masts were eventually replaced as ships of the period were modernised.

and battlecruiser was established, with two turrets fore and aft, the inner one raised to fire over the top of the outer one.

Tests had been made using the armoured turret of an old monitor to establish that blast effects from the superfiring turret would not be detrimental to the lower one. With all the big guns on the centreline, the balancing of the ship was simplified and the magazines could be located in areas of maximum safety. From the viewpoint of gunnery control – by now a vital aspect of battle management – all the main armament could be given instructions at once when firing salvoes on the beam.

The arc of fire was around 270 degrees. The 305mm (12in) guns had 13.72m (45ft) barrels weighing 47.85 tonnes (52.75 tons) and fired 390kg (870lb) shells with an extreme range of 18,000m (20,000yd) at their maximum elevation of 15 degrees. The rate of fire was three rounds a minute at peak performance. The secondary armament of 22 76mm (3in) guns was intended for defence against torpedo boats but was altered in 1917, with 14 76mm (3in)

South Carolina and _Michigan_ set the pattern of pre-1914 US battleship silhouettes with their gun arrangement, basket masts and tall, capped funnels.

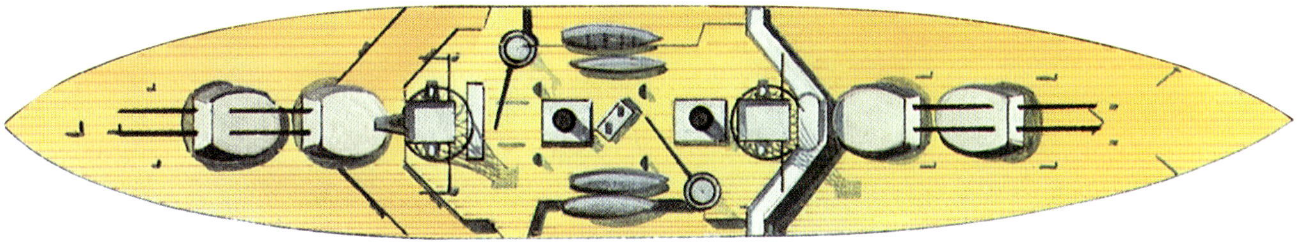

The deck plan clearly shows the simplicity and balance of the superfiring arrangement. The arc of fire for all guns was around 270 degrees.

guns, two 76mm (3in) AA guns and four 1-pounder guns. The AA guns were mounted on top of the derrick posts instead of torpedo searchlights, reflecting the fact that the airplane was the new threat to large warships.

Officers' quarters

Another consequence of the weight limitation was a shortening of the stern section from the preferred design, and the sacrificing of a deck level compared with the previous *Connecticut* class. American officers were normally accommodated at the stern (as in the Royal Navy) but in these ships their quarters were moved to the midships superstructure. Despite the weight restriction, *South Carolina* was well-armoured. The belt armour was thickest at 305mm (12in) where it protected the magazines of the main guns, and 279mm (11in) alongside the boiler and engine rooms, tapering to 203mm (8in) and closed off at each end by transverse 203mm (8in) bulkheads. Below the waterline the armouring was 254mm (10in). There was no longitudinal bulkhead dividing the ship, unlike French and British practice.

Turbine propulsion had been considered but price considerations prevailed and *South Carolina* had reciprocating engines that drove it at 34.26km/h (18.5 knots), 4.6km/h (2.5 knots) slower than HMS *Dreadnought*. But the US Navy command considered speed to be less vital than firepower and armour protection, at least until long-range German U-boats began to reach American coastal waters. The low freeboard of 3m (10ft), especially at the quarterdeck, made them wet ships and they had a notable tendency to roll in rough seas, though not to such a dangerous extent as the contemporary German *Nassau* class, and bilge keels were not applied.

South Carolina was deployed to the Atlantic Fleet, based at Key West and Guantanamo. After a refit at Philadelphia, completed in January 1917, it returned to Guantanamo. From the outbreak of war in April 1917 it was used as a gunnery training ship in Chesapeake Bay. In September 1918 it had to return from convoy escort duty for mechanical repairs. After the Armistice, from February to July 1919 it made four Atlantic crossings to repatriate American troops from France. With *Michigan* it made a training cruise to Honolulu in 1920 and went on further training and goodwill cruises until paid off at Philadelphia on 15 December 1921. It was stricken from the list on 13

Specification

Dimensions	Length 137.9m (452ft 8in), Beam 24.5m (80ft 5in), Draught 7.49m (24ft 7in), Displacement 14,515 tonnes (16,000 tons); 16,238 tonnes (17,900 tons) full load
Propulsion	12 Babcock & Wilcox boilers, 2 vertical 4-cylinder triple-acting engines developing 12,304kW (16,500hp), 2 screws
Armament	8 305mm (12in) guns, 22 88mm (3in) QF guns, 2 underwater 533mm (21in) torpedo tubes
Armour	Belt 305–228mm (12–9in), Casemates and barbettes 254–203mm (10–8in), Turrets 304mm (12in), Deck 38mm (1.5in), Conning tower 305mm (12in)
Range	9260km (5000nm) at 10 knots
Speed	18.5 knots
Complement	869

Lion (1912)

The three ships of the *Lion* class were the first battlecruisers to carry 342mm (13.5in) guns, and were the largest and fastest capital ships yet built; they were also the most expensive. But they had serious defects.

The British Navy adopted the superfiring turret arrangement in HMS *Neptune* (commissioned November 1911), and with the *Orion* class of battleships (commissioned January 1912) introduced the 'super-dreadnought' with 343mm (13.5in) guns. These aspects were combined in the *Lion* class battlecruisers. Britain had introduced the battlecruiser, as an enlarged version of the armoured cruiser, with HMS *Invincible*, commissioned in 1908. Like HMS *Dreadnought*, it was a project begun and driven by Admiral Lord Fisher who believed it to be tactically and strategically superior to the battleship.

Lion was the third class of battlecruiser to be introduced, laid down at Devonport Naval Dockyard on 29 November 1909, launched on 6 August 1910 and commissioned on 4 June 1912. Two others, *Princess Royal* and *Queen Mary*, completed the class. Each ship cost in excess of

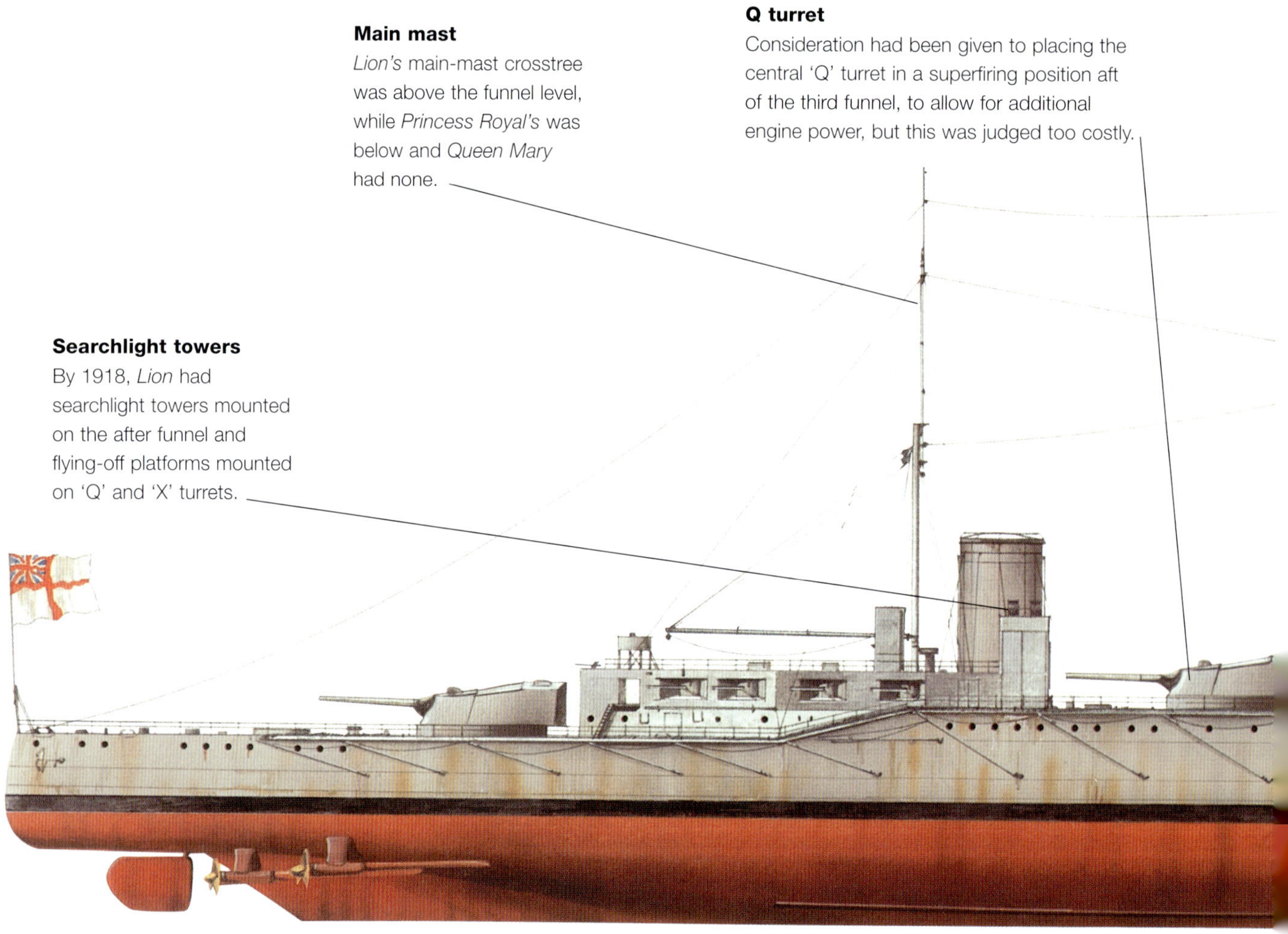

£2,000,000. The German Navy had responded rapidly to the implicit challenge: SMS *Moltke* was commissioned in September 1911 while *Lion* was still fitting out, and *Derfflinger* was laid down in January 1912. *Moltke* carried 10 208mm (11.1in) guns.

Design errors

Lion's design followed that of *Dreadnought* and *Orion* by having the fore-funnel placed in front of the mast. As a result sparks, smoke and heat made the masthead installations often uninhabitable. The bridge, placed on top of the conning tower, suffered similarly. In 1912 the original tripod mast was replaced by a single pole mast with a light spotting top, and the funnel was moved behind it, though

Larger than the contemporary German battlecruiser *Seydlitz*, and only half a knot faster, *Lion*'s armour accounted for 23 per cent of displacement, while *Seydlitz*'s was 31 per cent.

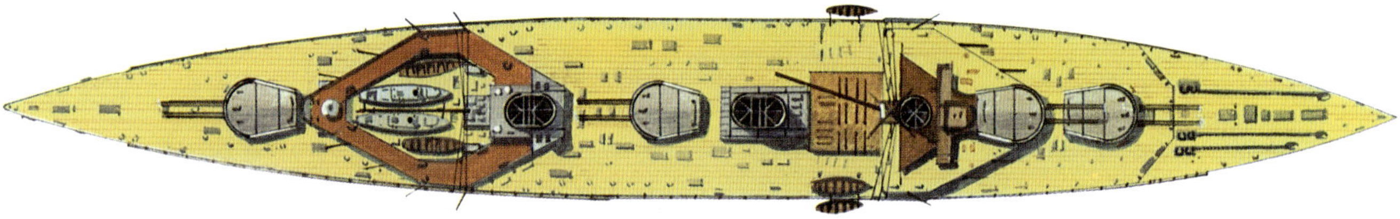

The deck plan shows how much of *Lion*'s 213.4m (700ft) length needed protection because of the spacing of the 342mm (13.5 in) turrets.

still very close. The second and third funnels were heightened to be uniform with the fore-funnel.

Although the 'A' and 'B' turrets were set in a superfiring arrangement, there was only a single aft 'Y' turret, with a central 'Q' turret placed between the second and third funnels, separating the boiler rooms below. Rangefinders were located in 'B' and 'Y' turrets and in the conning tower, with the fire-control position (this was later transferred to the mast, which was fortified by struts to support it). The ships carried 16 102mm (4in) guns for anti-torpedo boat defence, their batteries aligned so as to have six firing ahead, eight abeam and four astern. Two 533mm (21in) torpedo tubes were below the waterline on either side of 'A' barbette. In 1917 searchlight supports were fixed on the mast and the after funnel.

Armour limitations

The most controversial aspect of the battlecruiser, certainly after the Battle of Jutland, was its relative lack of armour protection. Speed was the great requirement insisted upon, and greater speed meant greater length and more surface requiring protection. In fact the vulnerability of the *Lion* class was more due to insufficient understanding of the flash effects of a shell explosion and the countermeasures necessary, rather than of lack of armour as such. Belt

With a beam width only 0.12 per cent of their length, the three ships of the *Lion* class had a slender, elegant appearance. This is a pre-1914 view.

Battle of Dogger Bank

The Battle of the Dogger Bank was fought when, from intercepted radio messages, the British came out to forestall a German bombardment of coastal towns: Admiral Beatty's force of five battlecruisers – *Lion*, *Tiger*, *Princess Royal*, *New Zealand* and *Indomitable* – with seven light cruisers and 35 destroyers against Admiral Hipper's three battlecruisers, *Seydlitz*, *Moltke* and *Derfflinger*, one armoured cruiser, *Blücher*, four light cruisers and 18 torpedo boats.

Outnumbered, Hipper sought to avoid a battle but the faster British force pursued his ships in a stern chase and by 09:30 on 24 January they opened fire at long range. The disabling of *Lion*, with all electrical power lost, prevented Beatty from giving effective signals, and the German ships, except for *Blücher*, got away. Though indecisive, the battle was considered a British tactical victory and enhanced the reputation of the battlecruisers, despite the damage to *Lion* having revealed the British ships' lack of staying power under heavy gunfire.

armour was fitted up to main deck level for the first time, but the armoured deck was only 25.4mm (1in) thick, and the barbettes extending down into the hull had 76.2mm (3in) armour. Altogether, armour weight came to 5624 tonnes (6200 tons) or 23 per cent of the design displacement. By comparison the German *Moltke* had belt armour to a maximum of 270mm (10.6in) and an armoured deck of 50mm (2in).

Lion joined the 1st Cruiser Squadron on commissioning, then was flagship of the 1st Battlecruiser Squadron from January 1913. In the 1914–18 war it was flagship of the Battlecruiser Fleet, and gave long-range support at the Battle of Heligoland Bight on 28 August 1914. At the Battle of the Dogger Bank, the only fight exclusively involving battlecruisers on 24 January 1915, a shot from *Lion* knocked out the rear turret of Admiral Hipper's flagship *Seydlitz*, but *Lion* sustained 17 hits including two on the waterline, narrowly avoided flooding of the engine room, and fell out of the action. Towed back by *Indomitable*, it spent four months under repair.

At the Battle of Jutland, 31 May 1916, *Lion* took a direct hit on 'Q' turret, whose officer, Major Harvey, died as he flooded the magazines, saving the ship. But its sister ship *Queen Mary* was blown up, with *Indefatigable* and *Invincible*. *Lion* was repaired by 19 July and continued in North Sea operations until the end of the war. In 1921 it was decommissioned under the Washington Agreement and was sold for breaking up in January 1924.

Specification

Dimensions	Length 213.4m (700ft), Beam 27m (88ft 6in), Draught 9.9m (32ft 5in), Displacement 23,832 tonnes (26,270 tons); 26,925 tonnes (29,680 tons) full load
Propulsion	42 Yarrow water-tube boilers, Parsons direct-drive steam turbines developing 52,199kW (70,000hp), 4 screws
Armament	8 343mm (13.5in) guns, 16 102mm (4in) guns, 2 533mm (21in) torpedo tubes
Armour	Belt 229–102mm (9–4in), Bulkheads 102mm (4in), Barbettes 229–203mm (9–8in), Turrets 229mm (9in), Deck 64–25mm (2.5–1in), Conning tower 254mm (10in)
Range	10,390km (5610nm) at 10 knots
Speed	28 knots
Complement	997

Wyoming (1912)

One of the US Navy's longest-serving battleships, *Wyoming* served in both World Wars in both the Atlantic and Pacific Oceans, and went through a series of changes in appearance and function.

American battleship design was a matter of steady progress, with stepped increases in size and firepower. From the Florida class (completed 1911) onwards they were powered by steam turbines. *Wyoming* (BB32) was laid down at Cramp's yard, Philadelphia, on 9 February 1910, launched on 25 May 1911 and commissioned at New York Navy Yard in September 1912. Its sister ship, *Arkansas*, was commissioned in the same month.

Gun configuration

Wyoming and *Arkansas* carried 12 305mm (12in) guns in six turrets. The barrel length was 50 calibre and weighed 51 tonnes (56.3 tons), firing shells of 394kg (868lb) and with a range of 24,900m (27,230yd) when elevated to 150. The space required by these long guns gave the midships section something of a squeezed-up look, with the two funnels and the basket masts all set close together. Derrick

Turbine propulsion
A mixed turbine-reciprocating propulsion system had been planned, but as completed the ships were turbine-only.

Quick-firing guns
Wyoming and *Arkansas* were the last US battleships to be fitted with quick-firing 305mm (12in guns). The six-turret configuration was unique to these two.

Armoured decks
Two armoured decks were fitted above the main magazines and the engine rooms.

posts were placed between the funnels. Original secondary armament was 21 127mm (5in) quick-firing guns. There were two armoured decks above the heavy-gun magazines and the engine room. Four Parsons turbines took steam from 12 Babcock boilers fuelled by a mix of coal and oil. *Wyoming* was assigned to the Atlantic Fleet and operated from Hampton Roads, Virginia, on routine patrols and exercises as flagship from 30 December 1912. In 1913 it made a long-range cruise as far as Malta, returning on 15 December. Until April 1917 it remained with the Atlantic Fleet. From April to November of that year it was used for engineer training, then crossed the Atlantic with USS *New York, Delaware* and *Florida* as Battle Division 9, to join the

The preceding *Florida* class had carried five gun turrets, but *Wyoming* and *Arkansas* were unique among US battleships in having six main turrets, with a potential broadside of 4728kg (10,423lb) of high explosive shells.

Forward guns
The 130mm (5in) guns forward of 'A' turret were often awash and were removed in 1919, as was the stern-mounted 130mm (5in) gun.

Wyoming approaches Brooklyn Bridge in December 1912, on returning from its extended cruise to Europe and the Mediterranean Sea.

Specification

Dimensions	Length 171m (562ft), Beam 28.4m (93ft 2in), Draught 8.7m (28ft 6in), Displacement 23,587 tonnes (26,000 tons)
Propulsion	4 Parsons turbines, 12 Babcock coal/oil boilers developing 20,880kW (28,000hp), 4 screws
Armament	12 305mm (12in) guns, 21 130mm (5in) guns, 2 533mm (21in) torpedo tubes
Armour	Bulkhead 279–229mm (11–9in), Belt 279–127mm (11–5in), Decks 76–25mm (3–2in), Turrets 305mm (12in), Barbettes 279mm (11in), Funnels 165mm (6.5in), Conning tower 305mm (12in)
Range	14,800km (8000nm) at 10 knots
Speed	20.5 knots
Complement	1063

British Grand Fleet at Scapa Flow, serving with the 6th Battle Squadron. In June–July 1918 it acted as a guardship during the laying of the North Sea Barrage, and afterwards was an escort to the surrendered High Seas Fleet on its way to Scapa Flow.

In December 1918 *Wyoming* returned to the USA, and from 19 July 1919 until August 1921 was deployed in the Pacific, ranging from Puget Sound to Hawaii and Valparaiso, Chile, on a mixture of tactical and diplomatic missions. From 15 September 1919 to 20 April 1920 it underwent refit at Puget Sound Navy Yard. The mast tops were enclosed and additional platforms for searchlights installed; the searchlights on the derrick tops were replaced by 76mm (3in) AA gun mountings, and improved rangefinding gear was installed on the bridge and on the superfiring turrets.

Extensive overhaul

A major modernisation was undertaken at the Philadelphia Navy Yard in the autumn of 1926. New turbines were installed and four White-Forster oil-burning boilers replaced the originals. Power output was raised to 20,879kW (28,000hp) and the speed increased to 38km/h (20.5 knots). These changes made it possible to dispense with the after funnel. The aft basket mast was replaced by a tripod structure, and a catapult was installed to launch a floatplane off 'C' turret. As a result the ship took on a quite different appearance.

Considerable alteration was also made to the secondary guns, with the three most forward casemates removed on both sides and the QF guns on the upper deck moved to positions projecting beyond the beam of the ship and fitting on top of the torpedo bulges. The torpedo tubes were removed. Further major modifications were made at the Philadelphia Navy Yard from 1930, when in accordance with the warship reductions required by the London Naval Treaty of 1930, Wyoming was 'de-militarised' and converted to a training ship. The 'C', 'D' and 'E' turrets were removed, as was the side armour. In this form the ship, now designated AG17, carried out training cruises and amphibious assault exercises in both the Atlantic and Pacific Oceans, including a visit to Kiel, Germany, in June 1937. A further overhaul at the Norfolk Navy Yard, 16 October 1937 to 14 January 1938, did not involve significant changes.

Gunnery training ship

The ship's training role was maintained and greatly intensified after the Pearl Harbor attack of 7 December 1941 and the USA's entry into World War II, with AA defence training as the prime task. Between 12 January and 3 April 1944 the fore basket mast was removed and replaced by a pole, and the remaining heavy guns and turrets were taken out to make room for a wide range of anti-aircraft guns and fire-control equipment. In June–July 1945 further modifications were made at the New York Navy Yard, in preparation for a deployment with Composite Task Force 69, working on ways to deal with drone planes and kamikaze tactics. *Wyoming* was decommissioned on 1 August 1947 and struck on 16 September. In 1947–48 it was broken up at Newark, New Jersey.

'Chesapeake raider'

Perhaps the most useful period of *Wyoming*'s long service was its World War II stint as a gunnery training ship, primarily for anti-aircraft defence. With the original big guns eventually all removed, the old ship carried 10 130mm (5in) 38-calibre guns in two single and four twin-mount turrets capable of high-angle fire, plus an array of smaller AA guns and fire-control equipment. It was claimed that *Wyoming* fired off more shells than any other ship in the US Navy during these years, not in any war zone but all into the Chesapeake Bay training area, gaining the nickname of the 'Chesapeake Raider'. In total, around 35,000 men were trained on seven or eight different gun types, mostly being transferred to destroyers and carriers once they had gained their badges.

The original deck-plan. As can be seen from the picture above, of the ship in the Atlantic on 25 April 1945, the ultimate armament and layout were very different.

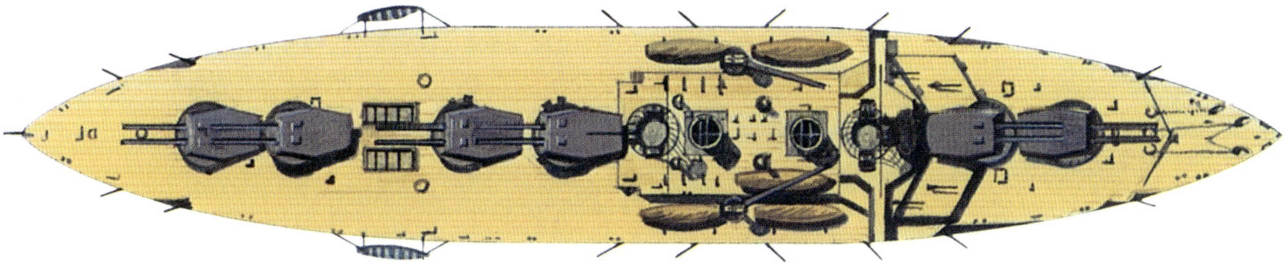

Derfflinger (1913)

Germany's battlecruisers, built in response to Britain's, were better armoured than their British counterparts, though not as fast. With the three ships of the *Derfflinger* class, the Imperial German Navy also first made use of the 305mm (12in) gun.

Plans for the class were completed in mid-1911. Their appearance was different from that of any previous German capital ship, with a flush deck rising gently towards a straight bow, and they were generally considered to be among the most handsome warships of their time. *Derfflinger* was laid down at the Blohm & Voss yard in Hamburg in January 1912, launched on 12 July 1913 (after sticking on the slipway for a month) and commissioned soon after the outbreak of war, on 1 September 1914. Total cost was 56 million Goldmarks. Its sister ships were *Lützow* (commissioned August 1915) and *Hindenburg* (May 1917).

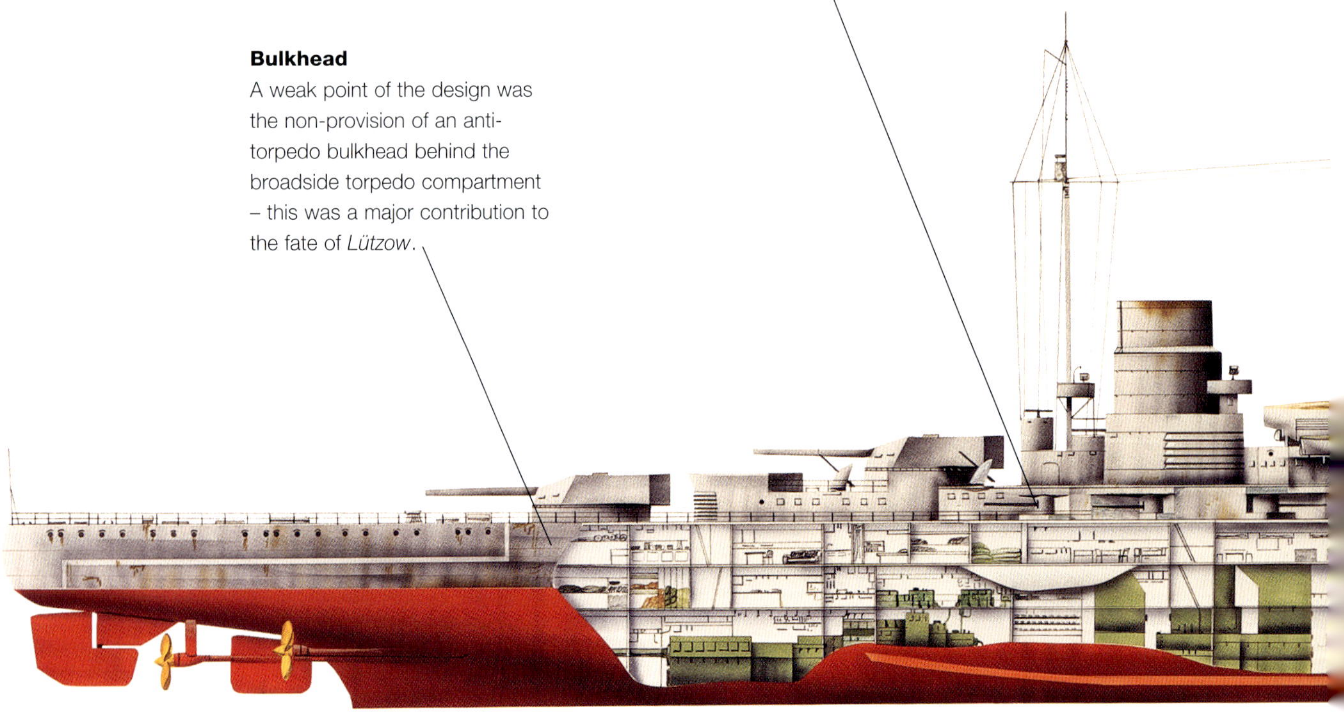

Weapons system

The heavy guns were placed fore and aft, with superfiring turrets. With a barrel length of 15.2m (50ft) and weight of 61 tonnes (67 tons), they fired shells of 405kg (893lb) for a distance of 16,200m (17,716yd) at an elevation of 13.5 degrees; this was later extended to 20,400m (22,310yd) and 16 degrees. Rate of fire was one round per minute. A secondary armament of 12 150mm (5.9in) quick-firing guns was fitted in casemates at deck level along the superstructure, and eight 88mm (3.4in) QF guns mounted at upper levels. Four of these were removed in 1916 and four AA guns of the same calibre mounted. Four underwater tubes for 500mm (19.6in) torpedoes were fitted, two on each beam, one each bow and stern.

Armour extended up to deck level between the main turrets and at the full length at waterline level, though there was a vulnerable area with no anti-torpedo bulkhead between the beam-mounted torpedo tubes. Overall, the

With its very slightly raked bow and long flush deck, *Derfflinger* had a very 'modern' look for a warship designed in 1911. Its boilers were partially oil-fired.

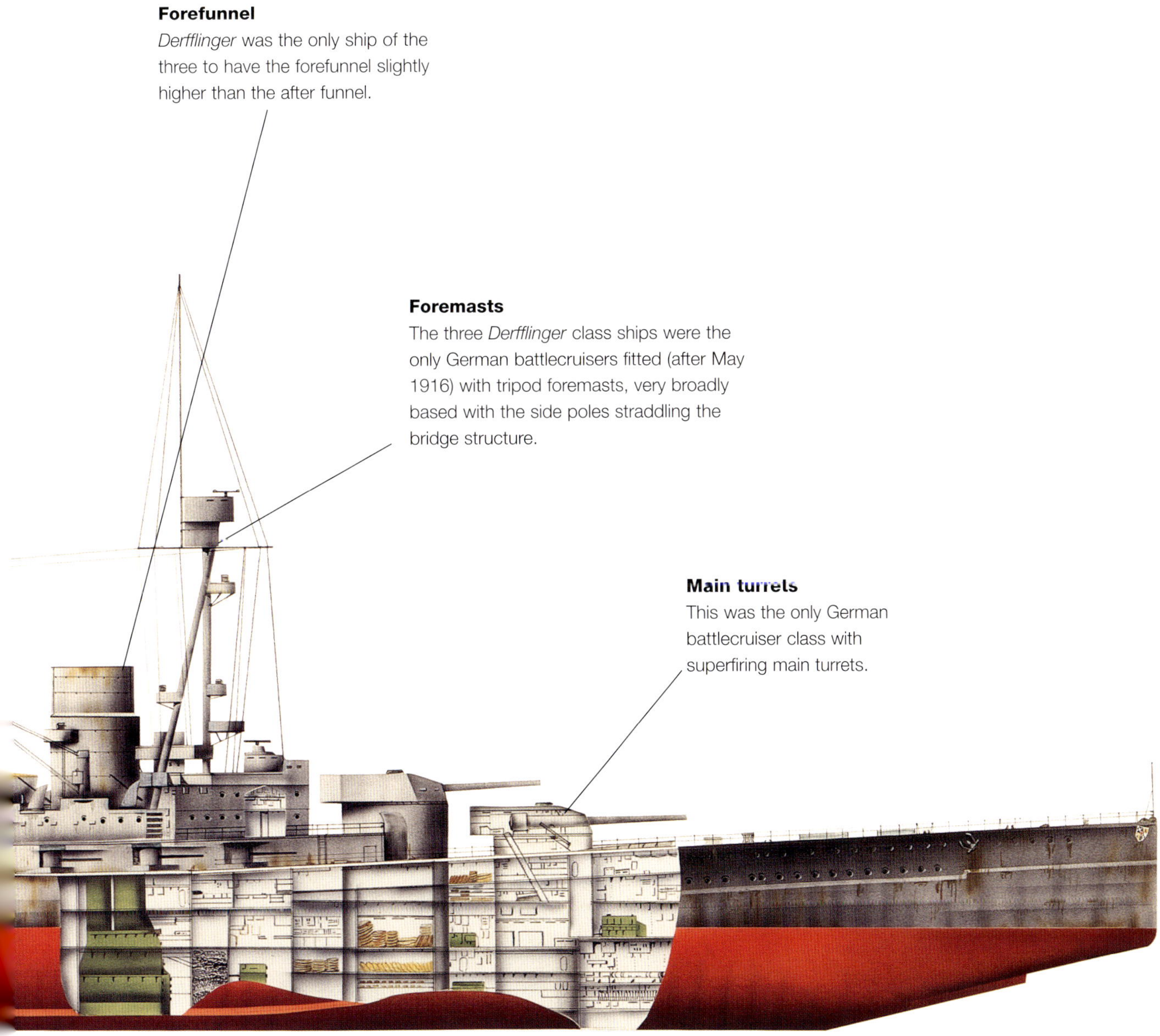

Derfflinger

This aerial view from 1917 shows the fine lines of *Derfflinger*'s hull design. Location is the Jade River, Kiel Naval Base.

German battlecruisers proved to have significantly more staying-power than their British counterparts.

A mix of coal and oil was used as fuel, with 14 twin naval boilers burning coal and four double-ended oil-fired boilers supplying steam to two turbines that drove four propellers. On *Derfflinger* the fore-funnel was mounted slightly higher than the after-funnel, and on both the caps rose above the outer casing. The ship fulfilled the battlecruiser's requirement for speed, making 47.8km/h (25.8 knots) on wartime trials. Power developed on that occasion was 57,121kW (76,600hp). Anti-rolling tanks were fitted to *Derfflinger* as an experiment, placed between the funnels in the central superstructure. Originally fitted with two pole-type masts, after Jutland a broad tripod mast replaced the forward pole mast with large viewing and direction-finding stations and a bridge-platform between its aft pillars.

Active service

Derfflinger was attached to Rear Admiral Franz von Hipper's First Reconnaissance Squadron. It took part in the bombardment of the English coastal town of

Specification

Dimensions	Length 210.4m (690ft 3in), Beam 29m (95ft 2in), Draught 9.2m (30ft 3in), Displacement 23,750 tonnes (26,180 tons); 27,857 tonnes (30,707 tons) full load
Propulsion	18 boilers, Parsons turbines, 4 screws
Armament	(1916) 8 305mm (12in) guns, 12 150mm (5.9in) guns, 4 88mm (3.4in) guns, 4 88mm (3.4in) AA guns, 4 500mm (19.6in) torpedo tubes
Armour	Side belt 300–150mm (12–6in), Bulkhead 250–100mm (9.8–3.9in), Conning tower forward 300mm (11.8in), aft 200mm (7.8in), Barbettes 260mm (10.2in), Turrets 270mm (11in), Deck 30mm (1.2in), Funnels 165mm (6.5in), Conning tower 305mm (12in)
Range	9816km (5300nm) at 14 knots
Speed	26.5 knots
Complement	1112–1182

Scarborough on 16 December 1914, an operation in which it was hoped to draw units of the British Grand Fleet into battle. It was at the Battle of the Dogger Bank, 24 January 1915, where it took a shell hit that caused localised flooding in protective coal bunkers. In August 1915 it was in action in the Gulf of Riga against Russian ships.

Back in the North Sea it made a sortie on 5–6 March 1916 into the 'Hoofden', the southern North Sea off the Dutch coast; it then joined in bombarding Yarmouth and Lowestoft on 24 April. It was with the High Seas Fleet at Jutland on 31 May. Hipper's battlecruisers were first to open fire in the battle and were heavily engaged until the end. With the battlecruiser *Seydlitz*, *Derfflinger* sank the British battleship *Queen Mary*, and, with *Lützow*, sank HMS *Invincible*.

Derfflinger took a considerable pounding, with 17 heavy hits and four lesser ones, and by the end of the battle it had only two guns still operational. But it returned to Wilhelmshaven naval base under its own steam, with around 2721 tonnes (3000 tons) of water inside the hull. Repairs were completed at Kiel by mid-October, but it remained in port except for a sortie into the northern North Sea, with *Hindenburg* and other ships, on 23 April 1918 in the hope of encountering a squadron of the Grand Fleet on convoy escort.

After the German surrender it went to Scapa Flow, 24 November 1918, and was scuttled there on Admiral Reuter's orders with the other German ships on 21 June 1919. Raised in 1939, it remained upside-down until 1948, when it was towed to Faslane for breaking up.

Battle of Jutland

The Battle of Jutland, fought on 31 May–1 June 1916, was the only large-scale sea battle of World War I. It began as an encounter between battlecruisers, with the German battlecruisers of Hipper's 1st Reconnaissance Squadron first to encounter the British ships of Beatty's 1st and 2nd Squadrons, before the battleships of the High Seas Fleet and Grand Fleet came within range of one another. *Derfflinger* was in action from around 15:38 on 31 May until the last moments of the battle in the early hours of 1 June, firing 385 shells from the main guns, 235 rounds from the secondary armament and a single torpedo. In total 157 of the crew were killed and 26 wounded. By 03:00 on 1 June, Hipper had to report that his squadron was no longer in fit condition to fight, and Admiral Scheer, commander in chief, ordered him to return to harbour with his surviving ships. *Derfflinger* was hit 17 times by heavy caliber shells and nine times by secondary guns (see picture).

Royal Oak (1914)

One of a class of five battleships intended for fighting in line with the main fleet, *Royal Oak* was engaged at Jutland. It was the first capital ship to be sunk in World War II, torpedoed at anchor in Scapa Flow.

Eight ships were originally to form the *Revenge* class, but only five were built. *Royal Oak* was laid down at Devonport Naval Dockyard on 15 January 1914, launched on 29 April 1915 and completed in May 1916. Its cost was £2,468,269.

The 381mm (15in) guns were the same as on the *Queen Elizabeth* class. Secondary armament was 14 152mm (6in) guns: the last time a main-deck level battery was included on a Royal Navy battleship, though they were set further back from the bows than on previous ships to minimise drenching by waves and spray. It had two 76mm (3in) AA guns and four 3-pounder guns. Four underwater 533mm (21in) torpedo tubes were fitted, with 20 torpedoes. These were removed by 1930, but in 1934–35 four torpedo tubes were fitted in the bow above the waterline, pointing abeam on each side.

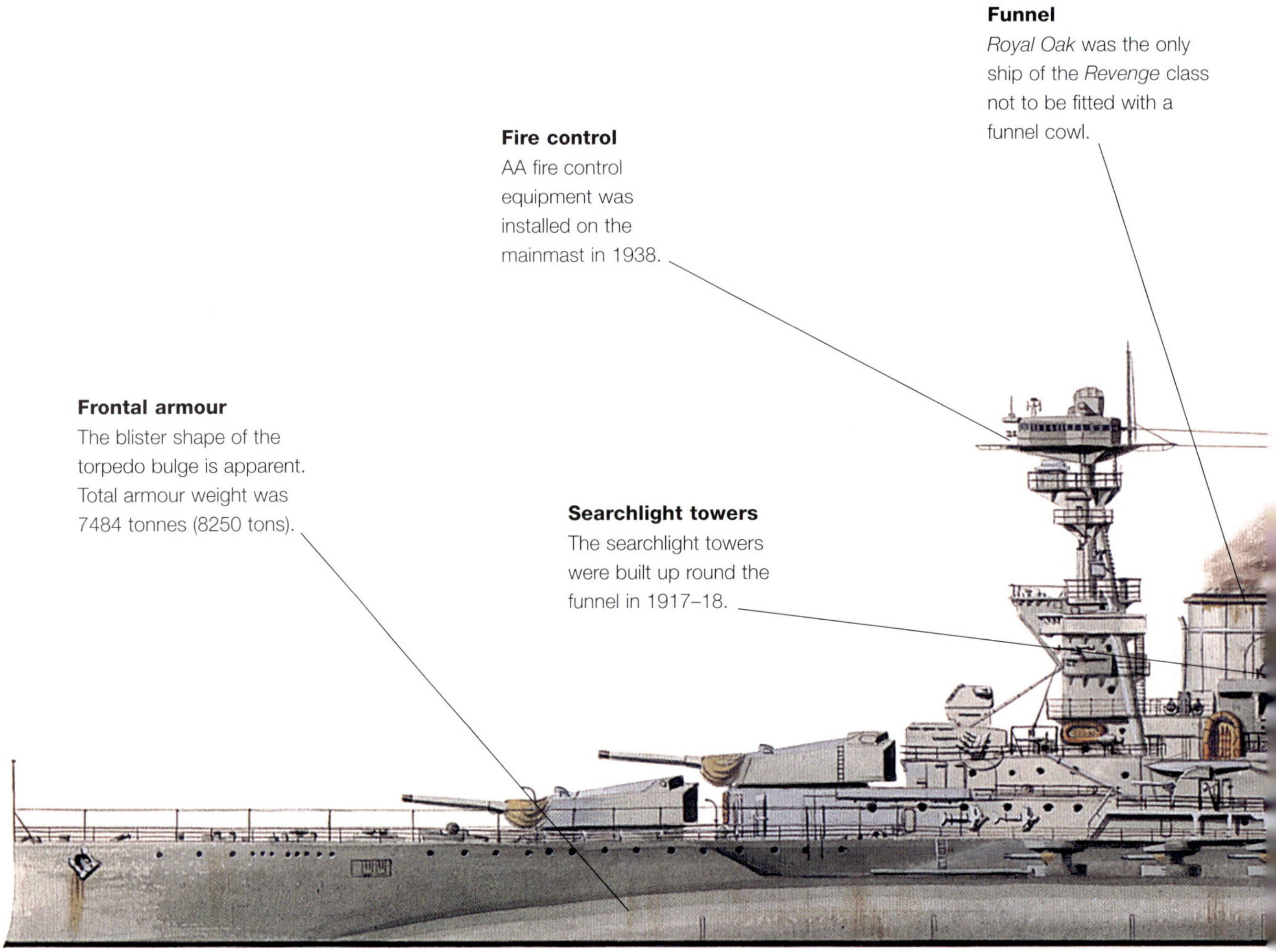

Speed improvement

Mixed coal-oil firing of the boilers was in the original plan, as high speed was not a requirement, but when Admiral Fisher was re-appointed as First Sea Lord in 1914 he demanded a change to oil only, giving an additional 3.7km/h (two knots) of speed. With three boiler rooms next to one another, the flues were trunked into a single massive funnel, and searchlight positions were built up about the after casing of it. By 1939 *Royal Oak* was the only ship in the class not to have a black-painted peaked cowl on the funnel-top.

Beginning with HMS *Ramillies*, the class were the first battleships to be fitted with anti-torpedo bulges, or 'blisters', adding 4m (13ft) to the beam, and in *Royal Oak* the bulges were heightened in 1927, almost reaching the battery deck. As on other RN battleships, the original secondary armament was progressively reduced and additional AA guns fitted. Between 1917 and 1923 all ships of the class had aircraft take-off platforms fitted to the superfiring turrets; and in 1934–1935 *Royal Oak* had a catapult mounted on the 'X' turret, and an aircraft crane was placed on the port side of the mainmast. Other modernisations made at Devonport at this time were a redesigned bridge structure, improved wireless communications and improvements to AA defences, including the mounting of an eight-barrelled 2-pounder pom-pom on each side of the funnel, along with two quadruple-mount 18mm (0.7in) machine guns on each side of the conning tower.

Up to 1938 the mainmast was of pole type, supporting a derrick, but when gunnery control instrumentation was installed at the crosstrees it was fitted with additional supports as a tripod. By this time, too, additions to the tower platforms and their combining into a single housing

Two ships of the *Revenge* class were ultimately built as the battlecruisers *Renown* and *Repulse* (1916). Of similar dimensions, they had six 381mm (15in) guns and turbines developing a maximum 93,958kW (126,000hp).

Secondary battery
This was the last RN battleship class with a main deck secondary battery on each side.

Aircraft launcher
This view shows *Royal Oak* as it looked in 1939, with aircraft launcher on 'X' turret and luffing crane.

Specification

Dimensions	Length 189m (620ft 6in), Beam (with bulge) 31.2m (102ft 2in), Draught 8.7m (28ft 6in), Displacement 25,991 tonnes (28,650 tons); 30,390 tonnes (33,500 tons) with bulges
Propulsion	18 Yarrow boilers, Parsons geared turbines developing 30,000kW (40,000hp), 4 screws
Armament	8 381mm (15in) Mk 1 guns, 12 152mm (6in) guns, 8 102mm (4in) guns, 4 533mm (21in) torpedo tubes
Armour	Belt 330mm (13in), Turret faces 330mm (13in), Barbettes 254mm (10in), Bulkhead 152mm (6in), Deck 51–25mm (2–1in)
Range	7400km (4200nm) at 10 knots
Speed	23 knots
Complement	997

made tower and foremast into an integrated structure. Nevertheless, in 1939 the extent of modernisation on board *Royal Oak* was quite limited. It still had the original engines, and its speed was no longer sufficient to maintain station with the fleet.

On commissioning, *Royal Oak* joined the 1st Battle Squadron of the Grand Fleet, took part almost immediately in the Battle of Jutland, 31 May 1916, and remained with the Grand Fleet until 1919, when it moved to the 2nd Battle Squadron and the Atlantic Fleet until 1922. Between then and 1924 it underwent a refit, then rejoined the 2nd Battle Squadron. From 1926 to 1934 it was flagship of the Mediterranean Fleet based at Malta. In a well-publicised episode, public disagreement between the Flag Officer on board and the captain and his chief officer led to their removal from their posts.

Torpedo attack

A further refit took place from May 1934 to August 1936, after which it returned to the 2nd Battle Squadron as

Torpedo bulges

Torpedo bulges had first been fitted to shallow-draught big-gun coastal monitors operating in U-boat-infested waters off the coast of Flanders in 1917. Flattening out fore and aft, it bulged outwards from the sides, intended to keep the centre of a detonation as far away as possible from the hull plating and to deflect the destructive energy upwards and downwards. Internally divided by a longitudinal watertight partition, the outer section of the bulge was empty and subdivided into watertight cells; the inner section was filled with water admitted through openings in the bottom. In normal loading, the bulge's top was just above the waterline. Speed was marginally reduced as a result (0.3 knots), but the staying power (and survival chances) of a ship in action were extended; the concept was soon copied by other navies. But *Royal Oak* succumbed to torpedoes fired from ahead and exploding against unprotected hull plates on the ship's bottom.

A veteran of the Battle of Jutland in 1916, HMS *Revenge* undertook escort duties in the early part of World War II, eventually being reduced to reserve status in 1943.

flagship with the Home Fleet. From 24 January 1937 it was back with the Mediterranean Fleet, at Gibraltar, and was involved in neutrality patrols off the Spanish coast during the Spanish Civil War, taking two minor hits from incidental firing. On 4 June 1937 it returned to Plymouth and was deployed with the Home Fleet. With the outbreak of war it was moved to the Scapa Flow anchorage, where it was torpedoed in a bravura operation on the night of 14 October 1939 by *U-47* (captained by Günther Prien). A first torpedo hit the bows, followed by a second salvo which scored three hits, exploding beneath the ship's bottom. It capsized and sank within 10 minutes with the loss of 833 lives.

Texas (1914)

USS *Texas* saw service in both World Wars in the Atlantic, the Mediterranean and the Pacific. The first US warship to launch an airplane, it still survives as a museum ship at La Porte in its name state.

Texas had pennant number BB35 and its sole class sister, *New York*, was BB34, though *Texas* was first to be laid down at Newport News on 17 April 1911. It was launched on 18 May 1912 and completed a month before *New York* in March 1914. Slightly larger than the preceding *Wyoming*, it carried heavier guns and cost $4,962,000 against *Wyoming*'s $3,856,000 (excluding guns in both cases). The first US battleship to carry 356mm (14in) guns, *Texas* had five twin turrets, three aft of the after-mast. Placing of the central turret shortened the superstructure as in other US battleships of the time, and the basket masts and twin funnels made a compact grouping.

American battleships were quite tubby compared to British ones; in both *Wyoming* and *Texas* the ratio of length to beam was 6/1, while in HMS *Lion* it was 7.9/1.

But *Texas* needed ample beam to fire its massive 10-gun salvo.

Barrel length was 13.7m (45ft), with a weight of 57 tonnes (63 tons), and firing shells of 635kg (1400lb). Each turret weighed 784 tonnes (864 tons). The guns elevated to 30 degrees and maximum range was 30,000m (32,800yd). Triple-expansion rather than turbine engines were installed due to disagreement between the naval authorities and the engine builders over the turbine specification. They were fired by 14 Babcock coal/oil boilers.

Block Island grounding

In 1914 there was a military-diplomatic crisis between the USA and Mexico, and *Texas* was first deployed in support of a US troop landing at Vera Cruz, and was then stationed off Tampico. From 1915 it was part of the

As commissioned in 1914, *Texas* had two basket masts, set closely fore and aft of two equal-size funnels, giving it a quite different profile to the post-1935 form shown here.

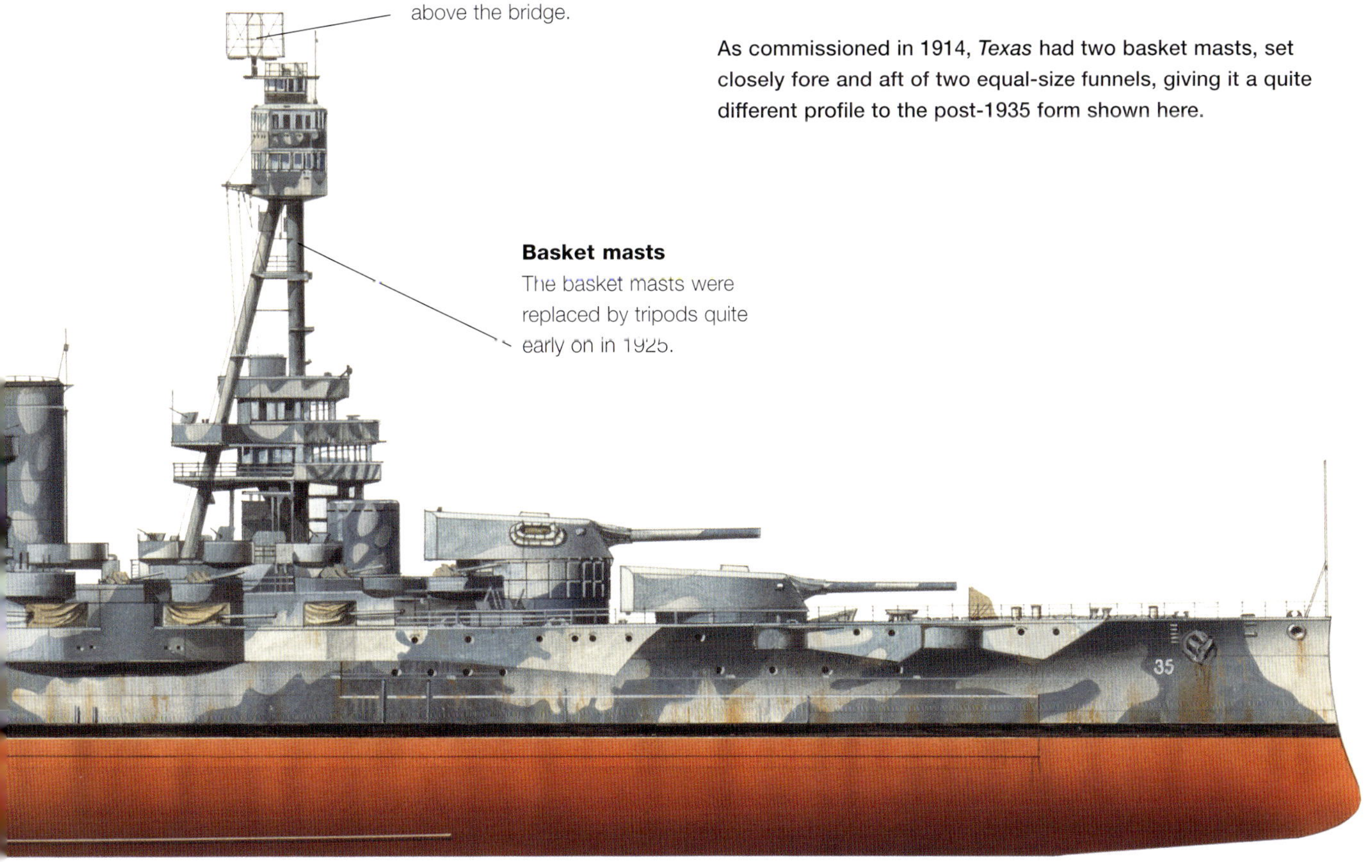

Radar aerial
In December 1938 *Texas* was fitted with a CXZ radar aerial above the bridge.

Basket masts
The basket masts were replaced by tripods quite early on in 1925.

Specification

Dimensions	Length 174.5m (572ft 7in), Beam 29m (95ft 3in), Draught 8.7m (28ft 5in), Displacement 24,494 tonnes (27,000 tons)
Propulsion	14 boilers, 2 vertical triple-expansion engines developing 20,954kW (28,100hp), 2 screws
Armament	10 356mm (14in) guns, 21 127mm (5in) guns, 4 533mm (21in) torpedo tubes
Armour	Belt 304–254mm (12–10in), Deck 76mm (3in), Turrets 356mm (14in), Barbettes 305mm (12in), Conning tower 305mm (12in)
Range	14,816km (8000nm) at 10 knots
Speed	21 knots
Complement	1530

A photograph of USS *Texas*, circa 1914, just back from foreign waters, its main guns and basket mast prominent.

Atlantic Fleet. As a unit of Battleship Division 9, it was due to cross the Atlantic to join with the British Grand Fleet, but ran aground at Block Island on 16 September 1917 and had to undergo repair at New York Navy Yard. It finally reached Scapa Flow in the Orkney Islands on 11 February 1918, from where it participated in convoy escort and North Sea patrols.

On return to the USA, *Texas* became the first US battleship to launch an aircraft on 19 March 1919. From that year until 1924 it was with the Pacific Fleet, then returned to Norfolk for a major modernisation between 1925 and 1927. Tripods replaced the basket masts, with the after one now set between 'C' and 'D' turrets. Six Bureau-Express oil-fired boilers replaced the previous arrangement, and the two funnels were replaced by a single one. A catapult was fitted on the central 'C' turret and cranes replaced the former derricks on each side of the funnel.

Compromised handling

Within the hull the torpedo tubes were removed (only one battleship is known to have fired a torpedo in combat in World War I), a new torpedo bulkhead was provided, along with a triple bottom in the midships section, and the horizontal armour was strengthened. As a result of the modifications the ship's displacement was increased by some 2721 tonnes (3000 tons) which had an adverse effect on handling, especially in rough conditions. In the 1930s it served first with the Atlantic then the Pacific fleet, and as a training ship. Further modifications at that time included the removal of the topmasts in 1934–35, and the addition of AA guns mounted on the tripod platforms. In 1938, a compact CXZ radar aerial was fitted above the bridge.

During 1940–41 *Texas* made patrols in the western Atlantic for the protection of US shipping against the belligerents, then, following American entry into the war, it operated as a convoy escort – a duty in which its comparatively low speed was not a problem. In November 1942 it provided artillery support for Allied landings in Morocco and Algeria, repeating the role in the Normandy landings of 1944 and also in southern France in the same year.

In 1942 an SG radar aerial was fitted to a platform on the foremast, and SK radar was added on the aftermast

in 1943. Also from 1943 six 20mm (0.79in) AA guns were mounted on the top of 'B' turret.

From late 1944 *Texas* was redeployed in the Pacific and saw action at Iwo Jima and Okinawa. Between 1941 and 1945 the ship fought in a total of 116 actions. It was the first of several battleships to be transferred by Act of Congress from the Navy to become a museum ship in 1948.

The deck plan shows the post-conversion layout, with single funnel, and tripod mainmast set aft of the centre turret.

Take-off platform

The 'take-off platform' was originally fitted to *Texas*'s 'B' turret during the 1919 refit at the New York Navy Yard. It was used to launch a British Sopwith Camel aircraft. Use of aircraft on capital ships began in 1914–18 and was initiated by the British. The key item was a floatplane that could be launched by catapult, land on water and be craned back on to the parent vessel.

Aircraft were used for reconnaissance and were more versatile than tethered balloons, which were also used. By the 1920s, the US Navy was using American-built aircraft, beginning with the Vought Corsair in 1927, then the Keystone OL9 in 1930, the Curtiss SOC 4 Seagull (1934), the Vought OS2 U-1 Kingfisher in 1941, the Curtiss SO3 C-2 Seamew in 1942 and the Curtiss SC-1 Seahawk in 1944. By late World War II, carrier-borne planes were available in such numbers that the British and Americans largely dispensed with auxiliary aircraft on battleships, though the Germans and Japanese used them to the end.

Queen Elizabeth (1915)

A 'super-dreadnought', class-leader of five battleships that served in both world wars. In World War I their speed and firepower set them apart from the rest of the Grand Fleet as a separate battle squadron.

Laid down at Portsmouth Dockyard on 27 October 1912, launched less than a year later (16 October 1913) and commissioned on 19 January 1915, *Queen Elizabeth*'s design was influenced by reports and rumours about the battleship designs of other powers, suggesting that the USA and Germany were planning higher-calibre naval guns. Consequently, 381mm (15in) guns were ordered, and though only eight were to be mounted it was felt that their weight of impact would more than make up for their number. At this time the Germany Navy's biggest guns were 304mm (12in) and most were 279mm (11in).

Six ships in the class had been planned but the sixth was not built (its name, *Agincourt*, was applied to a dreadnought carrying 14 304mm (12in) guns originally ordered by Brazil, sold while building to Turkey and confiscated by the British government in August 1914). *Queen Elizabeth*'s guns had

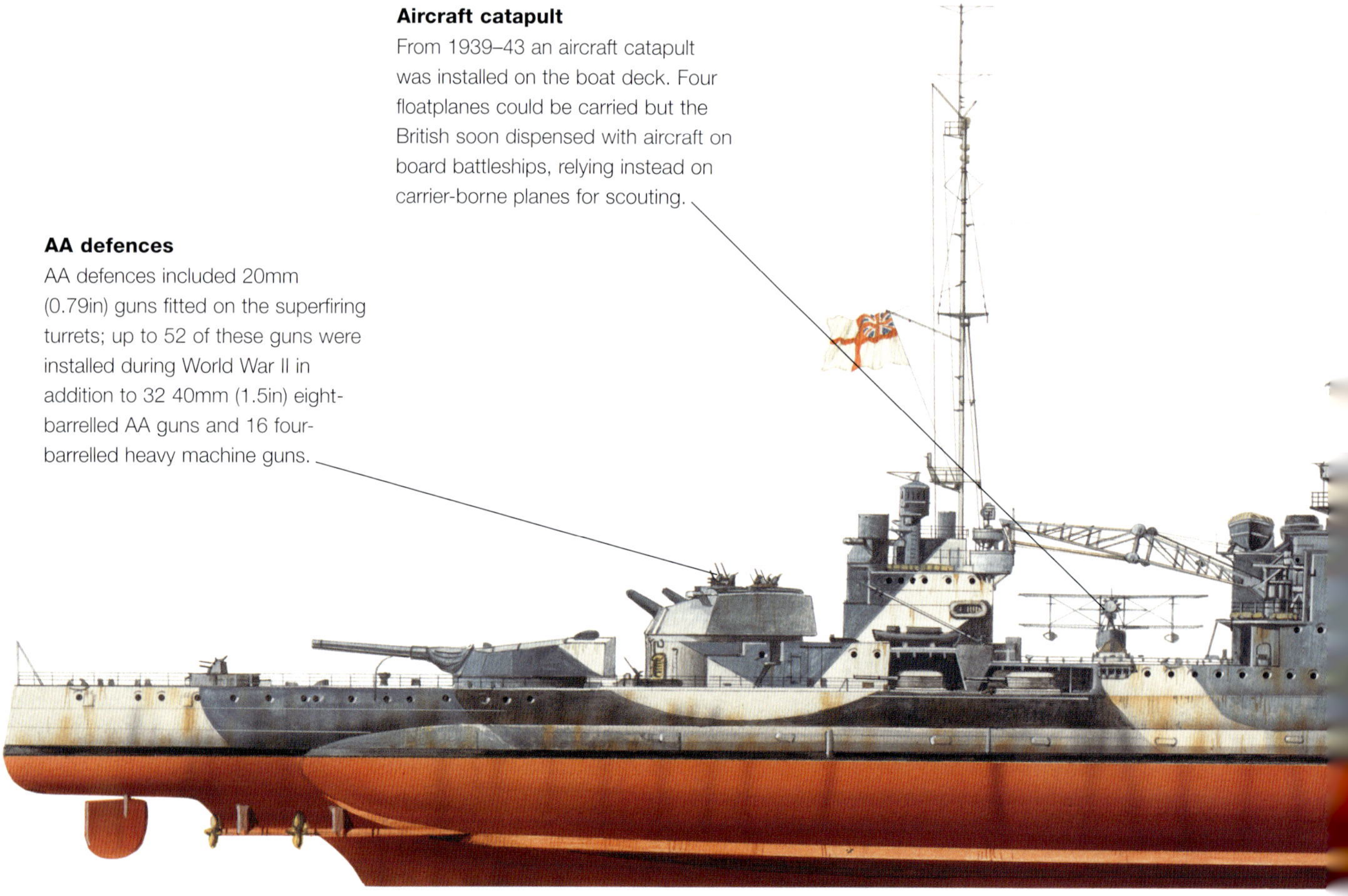

Aircraft catapult
From 1939–43 an aircraft catapult was installed on the boat deck. Four floatplanes could be carried but the British soon dispensed with aircraft on board battleships, relying instead on carrier-borne planes for scouting.

AA defences
AA defences included 20mm (0.79in) guns fitted on the superfiring turrets; up to 52 of these guns were installed during World War II in addition to 32 40mm (1.5in) eight-barrelled AA guns and 16 four-barrelled heavy machine guns.

barrels 12.8m (42ft) long, weighing 86.6 tonnes (95.5 tons) and firing shells of 875kg (1929lb), the heaviest naval shells used in World War I. The range was 32,000m (34,995yd) at an elevation of 30 degrees (the original maximum elevation was 20 degrees). Each turret weighed 928 tonnes (1023 tons) and was crewed by 75 men. Secondary armament consisted of 16 quick-firing 152mm (6in) guns mounted laterally in casemates. These were reduced to 12 by 1916, and two 102mm (4in) AA guns were installed.

Zeppelin threat

At this time, Zeppelin airships rather than airplanes were the main threat, though they also presented a more

Queen Elizabeth in 1943, with augmented AA armament and radar equipment after a refit in the United States. The aircraft and catapult were soon removed.

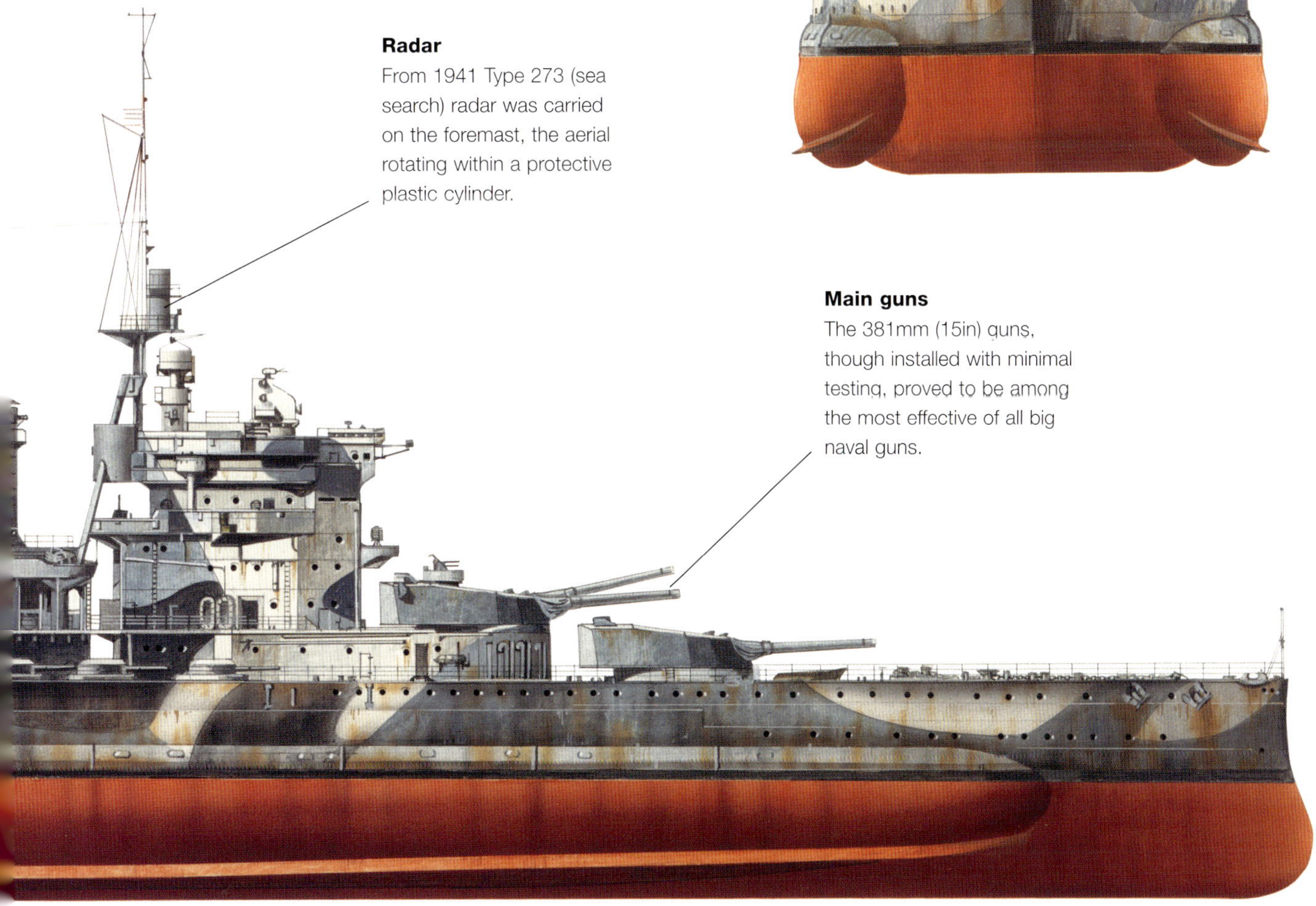

Queen Elizabeth

Queen Elizabeth around 1918. By this time torpedo booms and nets were no longer carried as a necessary part of a battleship's defensive equipment. Facing page: *Queen Elizabeth* around 1940.

Specification

Dimensions	Length 195.34m (640ft 11in), Beam 27.6m (90ft 6in), Draught 9.1m (30ft), Displacement 24,947 tonnes (27,500 tons); 29,995 tonnes (33,020 tons) full load
Propulsion	24 Babcock & Wilcox boilers, 4 Parsons geared turbines, developing 55,927kW (75,000hp), 4 screws
Armament	8 381mm (15in) guns, 16 152mm (6in) guns, 2 76mm (3in) AA guns, 4 533mm (21in) torpedo tubes
Armour	Belt 330–102mm (13–4in), Bulkheads 152–102mm (6–4in), Barbettes 254–102mm (10–4in), Turrets 330–127mm (13–5in), Deck upper 45–32mm (1.7–1.2in), Deck lower 25mm (1in) with 76mm (3in) over steering gear
Range	13,840km (7500nm) at 12.5 knots
Speed	23 knots
Complement	951

substantial target. Four beam-mounted underwater torpedo tubes were also installed. Power was supplied by 24 Babcock & Wilcox boilers in eight boiler rooms, serving Parsons turbines, high-pressure forward drive, low-pressure reverse drive and driving four propellers, plus a 'cruising' turbine that drove the outer screws at cruising speed. For the first time in any battleship only oil fuel was used, with a bunker capacity of 3084 tonnes (3400 tons).

Standard practice in most navies was for a battleship's armour to be of thickness equivalent to its gun calibre, but from *Dreadnought* (1905) onwards British designers had not followed this precept, in effect reducing armour in the interest of greater speed. On the *Queen Elizabeth* class, the maximum thickness was 330mm (13in) and this was applied only in a midships belt 1.2m (4ft) wide. Beyond this section, the armour tapered to 102mm (4in). The concept behind the requirement for speed was that a battleship squadron with a speed of some 7.4km/h (4 knots) greater than the main fleet would be able to pursue and outmanoeuvre an enemy fleet and force it into battle.

Dardanelles landings

On commissioning, *Queen Elizabeth* was immediately deployed to the Mediterranean Fleet and supported the Dardanelles landings from February to May 1915. Returning from that ill-fated venture, it joined the Grand Fleet after Jutland and was flagship from 1916 to 1920. From 1920 it was flagship of the Atlantic Fleet, then from July 1924 of the Mediterranean Fleet. In 1926–27 it underwent a major conversion following which it returned to the Mediterranean Fleet until 1929, when it was briefly flagship of the Atlantic Fleet, but within a year was back in the Mediterranean.

Between August 1937 and January 1941 it again had a major modernisation in Portsmouth, completed at Rosyth (see boxed text), and after a short attachment to the Home Fleet was sent again to the Mediterranean in May 1941. It was severely damaged at Alexandria on 19 December, along with its sister ship *Valiant*, by an intrepid attack from Italian midget submarines. After temporary patching-up it went to Norfolk, Virginia, for a complete repair, returning to the UK in June 1943 to again join the Home Fleet for a few months before being sent to the Far East, arriving at Trincomalee (Sri Lanka) on 28 January 1944 as flagship of the Eastern Fleet.

A period of intense activity followed, as escort and support ship for carrier operations and landings in the Indonesian archipelago, followed by a refit in Durban, South Africa, in October–November 1944. It then returned to the Indian Ocean for further service, including giving cover for landings on Ramree and action against Japanese-occupied Burma and Sumatra. On 12 July 1945 it returned to Britain and was placed on the reserve list in March 1946. Stricken in June 1948, it was scrapped in the same year.

Refitted warship

One rather anachronistic feature of *Queen Elizabeth* was a stern gallery, reminiscent of the old wooden men of war. Over the years, and especially between 1937–41, many alterations and additions were made to all ships in the class, substantially changing each vessel's profile. By 1942 only the hulls and the heavy guns remained unchanged, though the guns could now elevate to 30 degrees. New machinery was installed and a single large funnel mounted with a searchlight placement at its after edge.

The bridge structure was raised and enlarged to accommodate improved navigation and fire-control systems. Tripod masts were fitted fore and aft on *Queen Elizabeth*, an aircraft hangar and launching cranes were installed abaft the funnel, and a large number of 20mm (0.79in) AA guns were installed. Side armour alongside the 381mm (15in) magazines was strengthened to 102mm (4in). The number of 152mm (6in) guns was reduced to four on each side and the torpedo tubes were removed.

Fuso (1915)

The first 'super-dreadnought' of the Imperial Navy to be actually designed and built in Japan, *Fuso* was the world's largest battleship when commissioned. In original form it showed a mix of British and Japanese influences, but later modifications changed its appearance entirely, with an unmistakable profile.

Previous Japanese battleships had been built by British yards, though with much input from Japanese designers and technicians. *Fuso*'s design was partly based on the *Kongo*-class battlecruiser, but it was much more heavily armoured. The keel was laid at the Kure Naval Base on 11 March 1912. By the time it was launched on 8 November

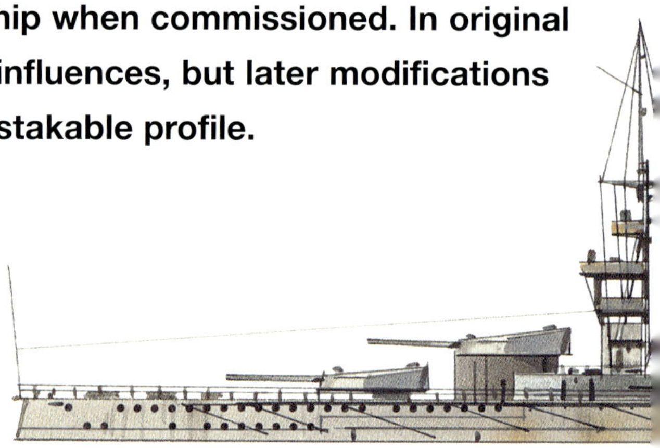

Turrets
The arrangement of No.3 and No.4 turrets was unusual, and the location of the central turret magazines in the same region of the ship as the oil bunkers may have contributed to its fate.

Guns
The original 356mm (14in) guns elevated from 00 to 300. Post-1933 elevation was between -50 and 430, with a maximum range of 35,450m (38,770yd). The Tetsukodan armour-piercing shell weighed 673.5kg (1485lb) and the high-explosive shell, 01-Shiki Tsujodan, 625kg (1378lb).

The two profiles, from 1915 (below) and 1933 (bottom), illustrate the remarkable change in *Fuso*'s appearance after modernisation. Completely new machinery was also installed.

Aircraft
Japanese capital ships carried Nakajima 90-11 (1927), Nakajima 90-11b (1933), Kawanishi 94-1 (1935), Kawanishi 95 (1938) or Mitsubishi F1 M (1938) floatplanes. All were biplanes.

Catapult
A Type Kure Shiki 2 Go 4 Gata catapult was mounted on No.3 turret in the 1930–33 modernisation.

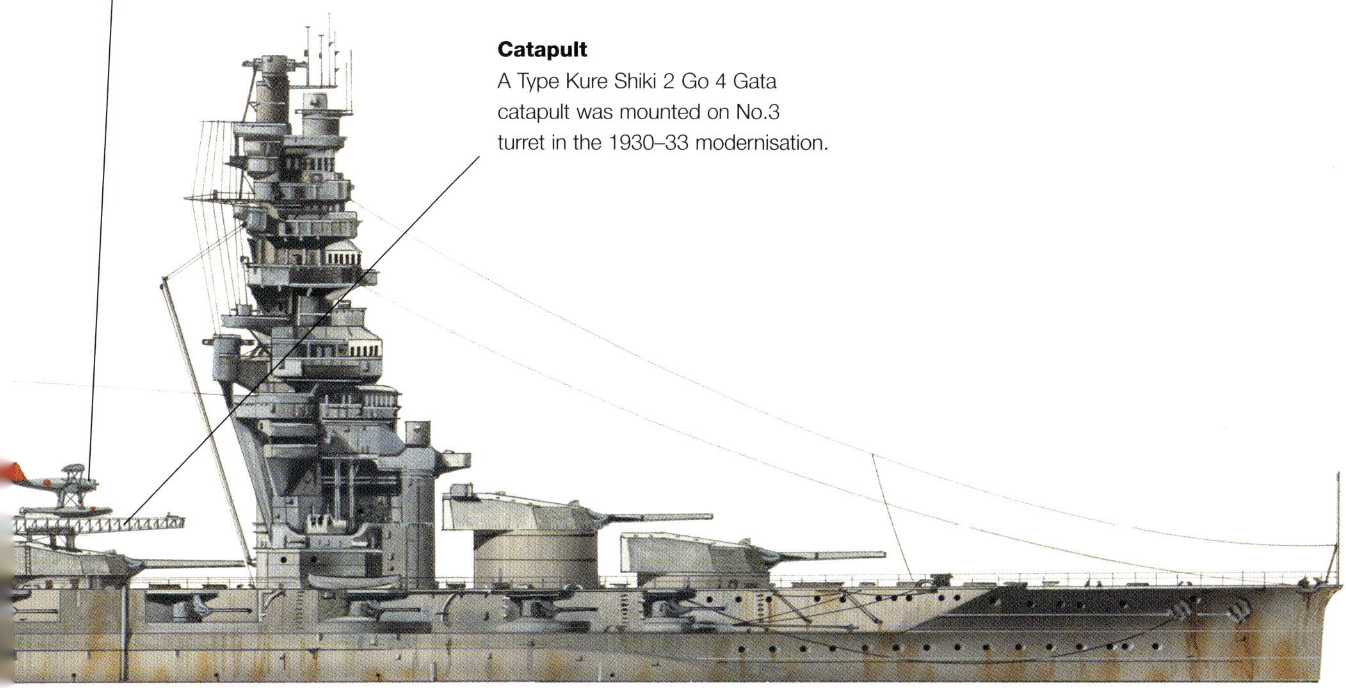

Fuso undergoing trials in 1915. It was one of very few battleships to have six heavy gun turrets.

1915, Japan was engaged on the Allied side against Germany. Attached to the First Squadron of the First Fleet, it was deployed off the Chinese coast, and, at the end of the war, alternated between first reserve status and patrols off China.

On 12 March 1930 a major modernisation programme began at Yokosuka Naval Yard. Completed in several phases, with service periods between, it included six new oil-fired Kampon boilers replacing the original eight double-ended and 16 single-ended Mijabara coal-fired boilers. More compact machinery reduced overall weight by 1152 tonnes (1270 tons) and enabled a single funnel to replace the two funnels, while almost doubling the original power output of 29,828kW (40,000hp).

A lofty 'pagoda' tower was erected, reaching 44m (144ft) above the waterline, with navigation bridges, lookout posts and fire-control stations; the foremast was discarded. With new 356mm (14in) guns, *Fuso* was effectively a new ship in the original hull, with speed boosted from 42.6km/h (23 knots) to 45.8km/h (24.75 knots). Originally, captive balloons had been used for aircraft spotting but by 1933 a catapult and three floatplanes were installed, two mounted on trolleys on the fore-deck and one on the catapult. A further modernisation programme began in 1937, lasting off and on until 1941, and was mostly concerned with the provision of AA defences and radar.

Doolittle Raid

In December 1941 *Fuso* acted as a support ship for the Pearl Harbor attack force, and in February 1942 it took part in the failed counter-attack on a US carrier force sending bombers over Tokyo (the 'Doolittle Raid'). After the Battle of Midway in May 1942 had demonstrated the power of aircraft against surface ships, there was a plan to convert *Fuso* and its sister ship *Yamashiro* to 'battleship carriers' with a short flight deck at the stern. In the end this change was applied only to the battleships *Ise* and *Hyuga*.

Type 21 radar was fitted to *Fuso* at Kure in June 1943, and Types 13 and 22 added in August. That month the ship participated in the reinforcement operation at Truk (Chuuk) in the Micronesian Archipelago. During the war years the AA batteries were built up and by 14 August 1944 *Fuso* had 95 25mm (1in) and 10 32mm (1.2in) guns in addition to the main and secondary armament.

Surigao Strait

On 20 October 1944, US forces landed on Leyte Island in the Philippines and the Japanese high command sought to concentrate its naval strength against the assault, resulting in a whole set of simultaneous battles, forming perhaps the largest sea fight in history. *Fuso*, captained by Rear-Admiral Ban Masami, and *Yamashiro* left their Brunei base on the 22nd, with the Third Squadron of the First Fleet under Admiral Nishimura. The 3rd Squadron was formed of *Yamashiro* as flagship, *Fuso*, the heavy cruiser *Mogami* and four destroyers, and its aim was to pass through the Surigao Strait, between the islands of Leyte and Mindanao, and link up with Admiral Kurita's First Fleet in Leyte Gulf.

The force was attacked by US aircraft on the 23rd, but fought them off without serious damage. A bomb penetrated *Fuso*'s quarter deck to explode in the wardroom, but the resulting fire was quickly put out. But positioned at the north end of the Strait was a formidable American battleforce, part of the 7th Fleet: the battleships *California*, *Tennessee*, *Mississippi*, *West Virginia*, *Maryland* and *Pennsylvania*, plus three heavy cruisers, two light cruisers and nine destroyers, commanded by Rear-Admiral J.B. Oldendorf. The forces engaged each other on the night of the 23rd–24th in what was to be the last 'classic' sea battle between heavy-gunned battleships. With American torpedo boats and destroyers behind, the Japanese squadron could only advance against devastating firepower. Both battleships were sunk.

Specification

Dimensions	Length 205m (672ft 6in), Beam 10.08m (33ft 1in), Draught 8.6m (28ft), Displacement 35,512 tonnes (39,145 tons) full load
Propulsion	6 Kampon boilers, 4 Kampon steam turbines developing 55,927kW (75,000hp), 4 shafts
Armament	12 356mm (14in) guns, 14 152mm (6in) guns (1938), 8 127mm (5in) guns, 16 132mm (5.2in) AA guns in quadruple mountings
Armour	Side belt 305–102mm (12–4in), Main turrets 297–114mm (11.7–4.5in), Casemate 152mm (6in), Deck 132–51mm (5.2–2in)
Aircraft	3 floatplanes
Range	21,854km (11,800nm) at 16 knots
Top speed	27.74 knots
Complement	1396

The fate of *Fuso*

The enforced radio silence of the Japanese ships, the darkness and the lack of survivors have left questions about the sequence of events in the battle of 24 October 1944, including whether *Fuso* or *Yamashiro* was the first to be sunk, and whether *Fuso* broke in two or, as has also been suggested, simply overturned, its hull flooded. The most likely outcome was that, battered by shells and torpedoes, at 03:45 *Fuso* suffered a tremendous explosion in the centre of the ship, where both ammunition magazines for the centre turrets and the oil bunkers were located, breaking the hull in two. Both parts remained afloat until the bow section was sunk by USS *Louisville*'s guns at 05:50. The stern section, blazing furiously, sank at around 06:50. All hands were lost, either on board, in the water or killed by locals if they reached the shore.

Nevada (1916)

The United States' first 'super-dreadnoughts', the two ships of the *Nevada* class incorporated a number of new features, including a revised armour arrangement, oil-firing and triple turrets.

Nevada, BB36, was laid down at Fore River Shipbuilding Co. at Quincy, Massachusetts, on 4 November 1912, launched on 11 July 1914 and commissioned on 11 March 1916. The US Navy by now was a formidable force and not short of innovative ideas. Having established the all centreline turret arrangement in 1906 (see *South Carolina*), it now established a new arrangement of protection.

Firing tests in 1911 against the old battleship *San Marcos* (ex-USS *Texas*) showed that light armour was of no value against heavy shells, and instead heavy armour, 343mm (13.5in) thick, was applied on 122m (400ft) of the 175.26m (575ft) waterline, with 203mm (8in) over the 18.9m (62ft) to the rudder-head. The barbettes also had 343mm (13.5in) protection as did the funnel uptakes, and the turrets had 457mm (18in) on the faces, 406mm (16in) on the sides and 254mm (10in) on the tops.

Other parts were not armour-protected, though internal bulkheads and an armoured deck were fitted. This, though

Life-boats
The mounting of Carley or similar-type life-boats on the sides of the main turrets was by now standard practice.

AA guns
In original form *Nevada* was equipped with only two AA guns of 76mm (3in) calibre. Reconstructed after Pearl Harbor, it bristled with AA defences.

Aircraft
Two catapults were fitted in 1927–30 for three Vought 02U-3 Corsair spotter planes.

not a new idea (see HMS *Inflexible*), was thoroughly worked out as the 'all or nothing' principle and became standard practice in battleship design after 1918. Weapons included 10 356mm (14in) guns, six of them in triple turrets. The guns were the same as those of USS *Texas*, though the triple sets were mounted in common cradles which meant that they could not be individually elevated. *Nevada* was powered by turbines while its sister ship *Oklahoma* had reciprocating engines; both were oil-fired, the first US battleships to be exclusively oil-burners.

Design alterations

Significant alterations were made during 1927–29. The secondary battery, originally in casemates below the forecastle deck, was raised to deck level, with the forward and aft guns removed, which made it more usable in rough conditions. The torpedo tubes were taken out and anti-torpedo bulges applied to the hull. Catapults were mounted on 'X' turret and on the stern 'fantail' deck, and tripods replaced the basket masts. These tripods were of massive construction, carrying three-

The profile shows *Nevada* as it looked in 1942, refitted after being damaged during the Japanese attack on Pearl Harbor, 26 years after its commissioning in 1916.

Funnel
The funnel was heightened by 2m (6ft 6in) in 1935; the distinctive angled smoke tube was added in 1942.

Mast
Nevada had three mast arrangements at different times. This shows the 1945 appearance after the topmast was heightened in 1943. Both masts carry Sra and SK radar aerials, the latter being added in 1943.

Guns
Nevada and *Oklahoma* were the first US battleships with triple turrets.

Specification

Dimensions	Length 583ft 4in (177.8m), Beam 95ft 3in (29m), Draught 28ft 5in (8.7m), Displacement 24,947 tonnes (27,500 tons); 26,217 tonnes (28,900 tons) full load
Propulsion	12 Yarrow boilers, 2 Curtis turbines plus 2 geared cruising turbines developing 19,761kW (26,500hp), 2 screws
Armament	10 356mm (14in) guns, 21 127mm (5in) guns, 2 76mm (3in) AA guns, 2 533mm (21in) torpedo tubes
Armour	Belt 343–203mm (13.5–8in), Bulkheads 38mm (1.5in), Deck upper 76mm (3in), Deck lower 38mm (1.5in), Funnel uptakes 305mm (12in), Barbettes 343mm (13.5in), Turrets 457–127mm (18–5in)
Range	18,520km (10,000nm) at 10 knots
Speed	20.5 knots
Complement	864

Nevada joined the Atlantic Fleet in 1916. From August to December 1918 it was based at Bantry Bay, Ireland, and with the end of hostilities returned to join the Atlantic Fleet. From 17 September 1927 to 26 November 1929 it underwent modernisation at Norfolk Navy Yard, and in early 1930 transferred to the Pacific Fleet. At Pearl Harbor on 7 December 1941 it was hit by an aerial torpedo and five bombs, and beached in shallow water. Refloated on 12 February 1942, it was repaired at Puget Sound and, with superstructure completely rebuilt, returned in early 1943 to Pacific operations at the Aleutian Islands, then from June returned to the Atlantic.

Normandy landings

Convoy escort duties were followed by support for the Normandy landings in June 1944, when it bombarded the

decked platforms with control, observation and signalling positions.

In the post-Pearl Harbor reconstruction, the appearance of *Nevada* was again drastically altered with a unique rearwards-tilted cap mounted on the funnel, the rear tripod removed and an array of SK and SRA radar aerials mounted on the forward tripod. The control tower was heightened, with Mk37 gunnery control equipment to control eight new 127mm (5in) twin-gun mounts.

In total 48 40mm (1.5in) and 27 20mm (0.79in) AA gun positions were installed, massed on the superstructure, which was redesigned to give them the widest possible firing arcs, and also fitted in emplacements on the bow and stern. Life-rafts were fitted on each side of the four turrets. As with many other US battleships an aircraft crane was mounted on the fantail.

***Nevada* sometime after its 1942 refit, with much-enhanced anti-aircraft defences in evidence. A plane is poised for take-off at the stern.**

Nevada's sister-ship *Oklahoma* was the last US warship to have triple expansion reciprocating machinery. Sunk at Pearl Harbor, it was formally decommissioned on 1 September 1944.

French port of Cherbourg and German tank formations inland. In August it was deployed to the Mediterranean to assist in landings on the southern French coast. On return to the USA later that year it had new gun barrels fitted and went again to the Pacific, in action at Iwo Jima and Okinawa between February and June 1945. Hit on 27 March 1945 by a kamikaze aircraft, it was only lightly damaged.

Decommissioned

Nevada was decommissioned at Pearl Harbor on 30 October 1945. In 1946 it was used as a target in the Bikini Atoll atomic bomb tests but remained afloat. Despite further use as a target ship, including shelling from USS *Iowa*'s 406mm (16in) guns, it survived until sunk by an aerial torpedo on 31 July 1948.

Oil firing

The transition to oil firing had a major influence on design. On earlier battleships, lateral coal bunkers had been considered as part of the defensive system, able to absorb shell or torpedo hits. This aspect was completely abandoned. *Nevada*'s oil fuel weighed 1848 tonnes (2037 tons), compared to *Texas*'s 2623 tonnes (2892 tons) of coal, and was carried in double-bottom tanks. Oil-fired boilers were more compact and lighter, enabling more weight to be given to armour and more space to be given to stores, accommodation and protective compartments.

With the whole propulsion plant taking up only 24m (78ft 9in) of a total waterline length of 175.3m (575ft 1in), *Nevada* and *Oklahoma* were the first US battleships to have a single funnel from the start. Refuelling was speeded up and refuelling at sea was much easier. While ships could be (and were) re-coaled at sea, it was a tricky and lengthy operation. Oil-fired ships also needed only half as many stokers.

Bayern (1916)

Bayern and the three others in its class were Germany's equivalent to the British _Queen Elizabeth_ class, with the same calibre and layout of heavy guns – but without the same turn of speed.

The plan of *Bayern* marked a break with the former German policy on guns. The smaller calibre of naval guns compared to that of the British battleships had been strongly criticised as well as strongly defended. While the move to 380mm (15in) calibre has been seen as a response to the same guns on the *Queen Elizabeth* class, it seems that the German decision was taken before the size of the new British gun was known. It was announced on 6 January 1912, before *Queen Elizabeth* was laid down and nearly two years before *Bayern* was laid down at Howaldtswerke, Kiel, on 20 September 1913. Launched on 18 February 1915, it was completed in March 1916. Its cost was around 50 million Goldmarks.

Main fleet service

Of its planned sister ships, only one, *Baden*, was completed before the end of World War I. Unlike the British *Queen Elizabeth*, whose gun-calibre they matched, they were

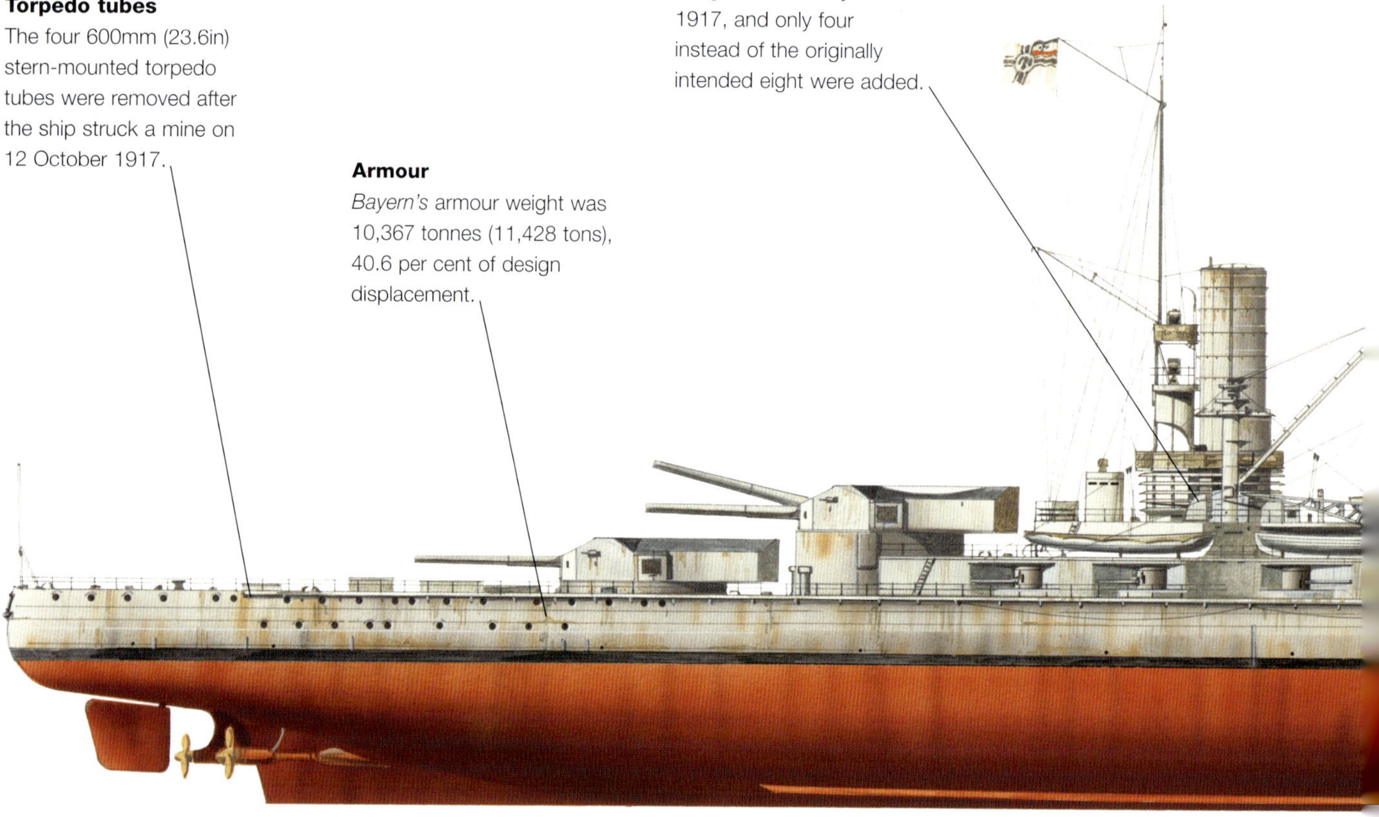

AA guns
The fitting of 88mm (3.4in) AA guns was delayed until 1917, and only four instead of the originally intended eight were added.

Torpedo tubes
The four 600mm (23.6in) stern-mounted torpedo tubes were removed after the ship struck a mine on 12 October 1917.

Armour
Bayern's armour weight was 10,367 tonnes (11,428 tons), 40.6 per cent of design displacement.

intended for main fleet service, not as a battle squadron. But Germany, with a numerically smaller fleet, had less reason to seize the initiative at sea, and the Marinamt could rely on the speed of the battlecruisers if it was necessary to cover a strategic retreat. Although of equivalent calibre, the

***Bayern* and *Baden* were the last additions to Germany's World War I battlefleet. It is interesting to compare *Bayern* with the British *Queen Elizabeth* class, to which in some respects it was superior, including those of armour protection and internal hull subdivision.**

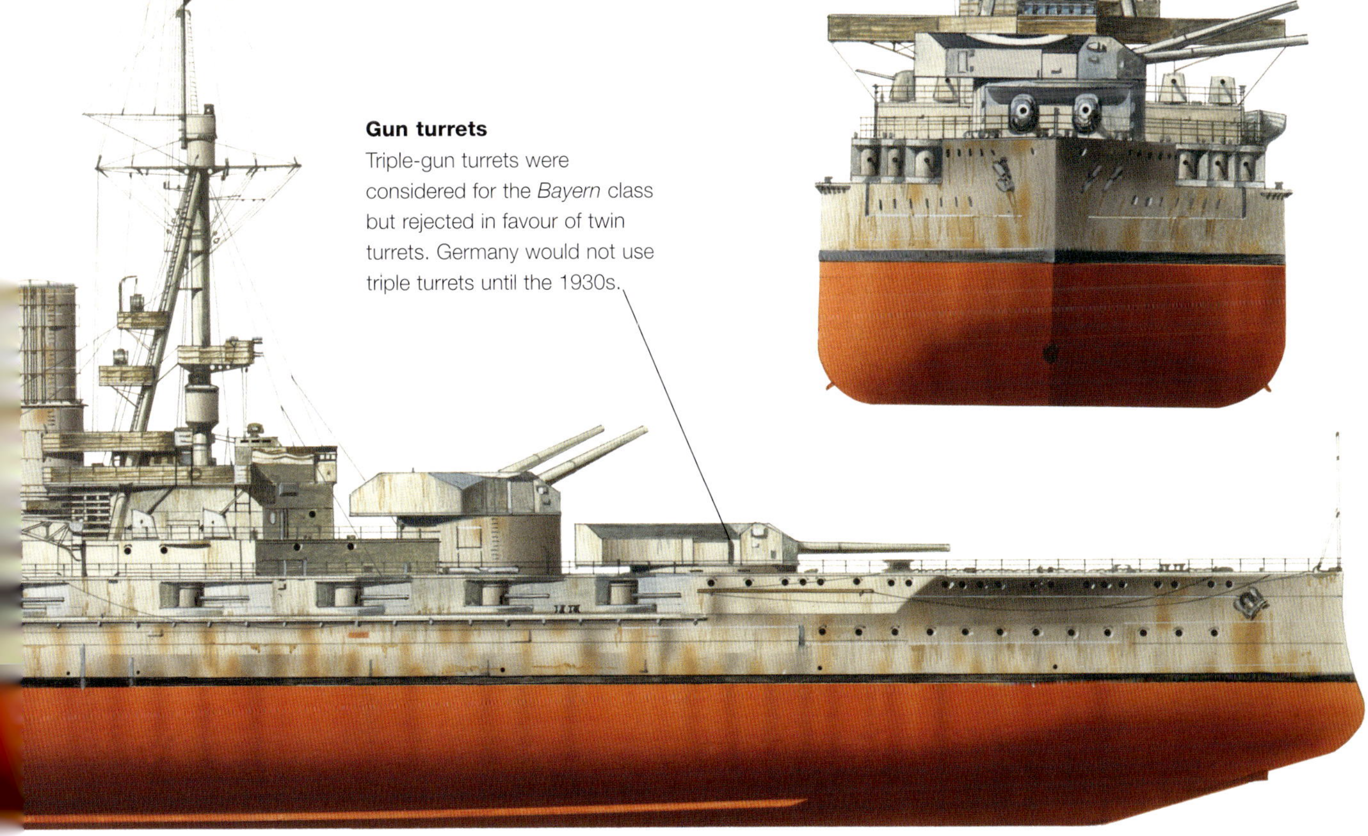

On tow to the scrap-yard in 1935, the capsized hull reveals the form of the propeller-shaft tunnels and the double rudder. The salvors have built a pumping-shed on 'top'.

German ships' eight 380mm (15in) guns were less effective than the British ones. With barrels 13.7m (45ft) long, they weighed 94 tonnes (103.3 tons) but fired a less heavy shell of 750kg (1653lb). Their range at first was 20,400m (22,310yd) but this was extended to 23,200m (25,372yd) when maximum elevation was raised from 16 to 20 degrees. They fired a round every 2.5 minutes, compared to two minutes for *Queen Elizabeth*.

Secondary armament was 16 150mm (5.9in) quick-firing guns arranged in traditional style in casemates, all at the same level, eight on each side. From 1917 four 88mm (3.5in) AA guns were carried. Originally, *Bayern* was fitted with five underwater 600mm (23.6in) torpedo tubes but four stern tubes were taken out in 1917, leaving a single bow tube.

When commissioned *Bayern* had a single mast, a tripod with five platforms and a tall topmast rising from the peak of the tripod. Two aerial poles and a flag gaff projected at an angle from behind the after funnel. These were replaced in 1917 by a relatively short pole mast stepped to the funnel casing. In total 11 coal-burning and three oil-burning naval boilers supplied steam to three Parsons turbines that drove the triple screws (the never-completed sister-ship *Sachsen* would have had six-cylinder two-stroke MAN diesels as well as two Parsons turbines). *Bayern*'s coal capacity was a maximum 3035 tonnes (3346 tons), plus 553 tonnes (610 tons) of oil. The difficulty of procuring sufficient oil supplies kept the German Navy chiefly reliant on coal while the British and Americans were turning to liquid fuel.

Inner defences

Four continuous longitudinal bulkheads formed the main inner defences. Coal bunkers 1.8m (6ft) wide ran along the sides between the outer gun turrets, separating the outer shell of the hull from the inner torpedo bulkhead. Transverse bulkheads were pierced for various kinds of piping including ventilation trunks, which reduced their protective power. Five 533mm (21in) sea valves were fitted on each side amidships in order to allow for localised flooding of

Deck plan: the balance of the ship's layout is clear. The rings on the superfiring turrets appear to have been applied to canvas covers.

compartments. To help trim the ship two further valves were placed aft and one forward.

After the war, British designers were able to closely examine *Baden*, and noted that in some respects, despite a greater emphasis on protection, the *Bayern* class was of inferior finish. The outer bottom plating was only 15.8mm (0.6in) compared to 25.4mm (1in) in HMS *Revenge*. Armour plating apart, British battleships were of more solid construction than the German ones. However, British battleships were built to range the world's oceans while the High Seas Fleet was intended for battle in the North Sea.

Though completed in March 1916, *Bayern* was still running trials during May and was not involved in the Battle of Jutland. On 19 August and 19 October 1916 it made unsuccessful sorties into the North Sea. In 1917 it was in action in the Baltic Sea against Russian units, and struck a mine on 12 October. During the necessary repairs, the various modifications noted above were made. It was part of the battle squadron that broke out on 23 April 1918, reaching as far as 58.97°N but failing to intercept British convoy escort ships. From 26 November 1918 it was interned at Scapa Flow, and scuttled there on 21 June 1919. Raised in September 1934, it was broken up at Inverkeithing in the following year.

***Bayern* capsizing on 21 June 1919. The four main turrets fell out of their mountings and still remain at the bottom of Scapa Flow.**

Specification

Dimensions	Length 180m (590ft 7in), Beam 30m (98ft 5in), Draught 9.4m (30ft 10in), Displacement 28,758 tonnes (31,700 tons)
Propulsion	14 marine boilers, 3 Schichau turbines developing 38,776kW (52,000hp), 3 screws
Armament	8 380mm (15in) guns, 16 152mm (6in) guns, 2 88mm (3.5in) guns, 5 600mm (24in) torpedo tubes
Armour	Belt 350–170mm (13.7–6.6in), Bulkheads 200–170mm (7.8–6.6in), Deck 150–30mm (5.9–1.1in), Barbettes 350mm (13.8in), Turrets 350–250mm (13.8–9.8in), Conning tower forward 350mm (13.8in); aft 170mm (6.7in)
Range	14,500km (7800nm) at 10 knots
Complement	1171

Courageous (1917)

Built for an invasion that never happened, and labelled as 'battlecruisers', the *Courageous* class were completely unsuited for the role. All three were converted to aircraft carriers and gave valuable service in that form.

Whether the 'battlecruiser' was really a viable concept will always be a source of debate. But there is no doubt that HMS *Courageous* and its two sister ships, *Furious* and *Glorious*, exhibited all the weaknesses and few of the strengths of the type. Planned for service in the Baltic Sea, to support a Russian invasion of northern Germany, they were first designated as 'large light cruisers' (funds for cruisers were available at the time, but not for additional capital ships), but later were officially listed as battlecruisers.

Courageous was laid down in May 1915 at the Elswick yard on the Tyne, launched on 5 February 1916 and completed in January 1917. *Glorious* was completed in the same month and *Furious* followed in July, their commissioning hustled along by the forceful Admiral Lord Fisher, 'father' of the project.

Lightly armoured

The hulls, far longer than any light cruiser's, were given only light armour, 76mm (3in) at its widest, though of up-to-date type. For the first time, resilient high-tensile steel was used, intended to cause HE shells to detonate prematurely on impact. Internal 37mm (1.5in) longitudinal bulkheads were fitted to absorb the resultant shocks. Integrated torpedo blisters were also fitted. One reason for the light armour

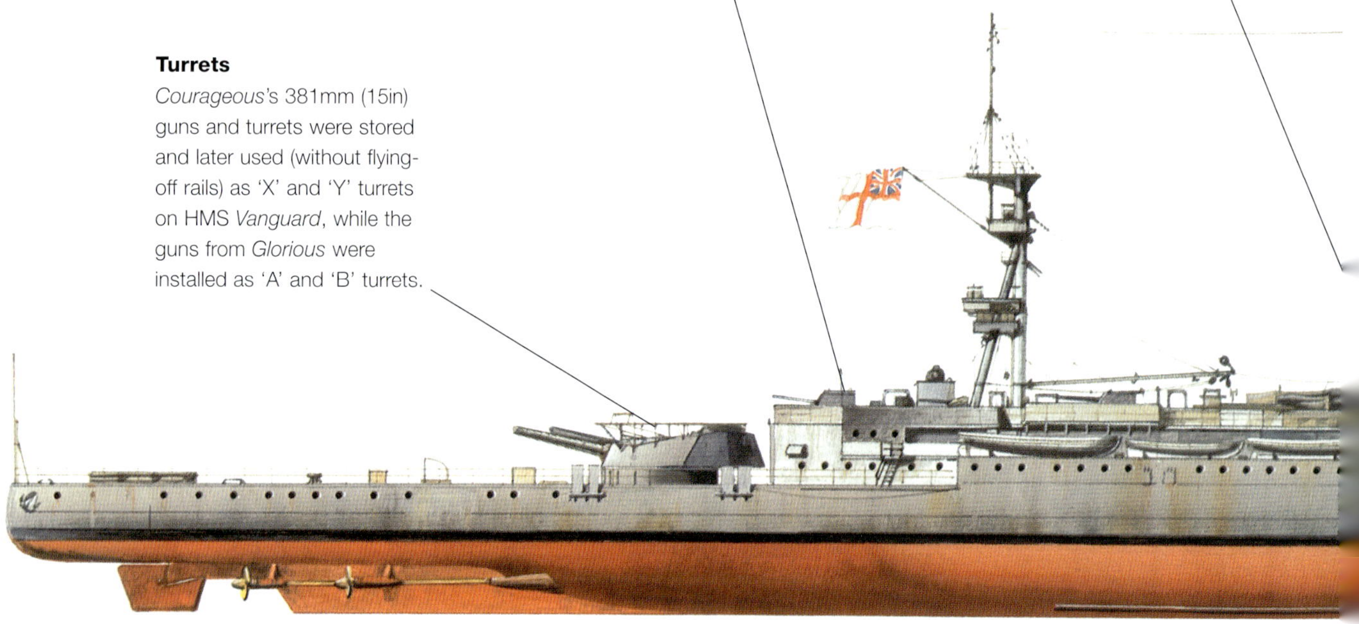

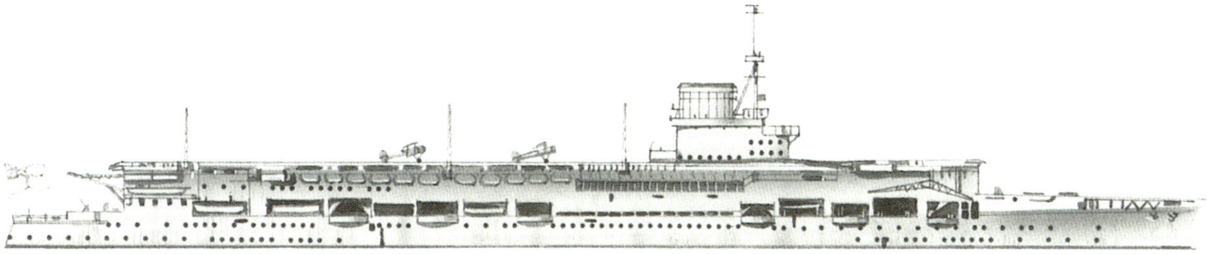

Above: *Courageous* in 1938. She was converted into an aircraft carrier in 1924–28, with an original complement of 48 aircraft. A leading warship historian described the *Courageous* trio as 'the largest tin-clad targets afloat'. Another writer called them 'Fisher's Folly'. But converted to aircraft carriers, they gave very effective service.

Below: *Courageous* in 1918 after fitting of the tripod mast. She also included flying-off platforms on both main turrets.

Engines
The best aspect of *Courageous* was the geared turbine engines which gave it a 30-knot speed.

Aircraft
Light planes could take off from the ship but not land, reducing their usefulness.

Courageous

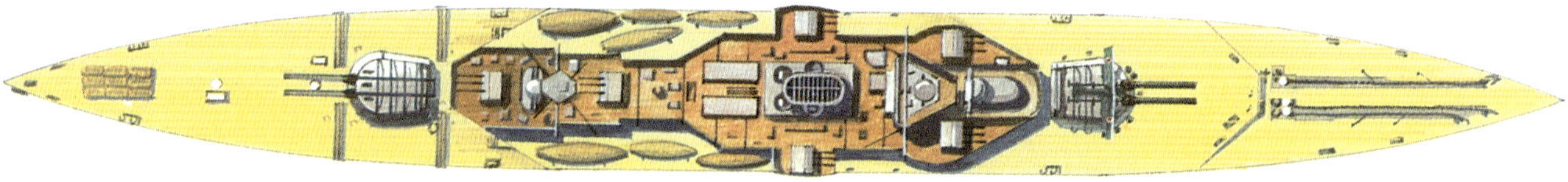

Deck plan: the original configuration of *Courageous* and *Glorious*. Later alterations would see mine-racks and chutes at the former's stern, and eight torpedo tubes on the latter's.

was that to operate in shallow Baltic waters, the draught was set at a maximum of 7.3m (24ft). This precluded heavy armour and to save further weight, the ships had only two main turrets. The guns were large-calibre 381mm (15in) in paired mounts, except for *Furious*, which for a short time had the largest gun mounted in any British warship, of 457mm (18in) calibre, set in a single aft turret. The trio did not lack speed, being powered by 18 Yarrow oil-fired boilers, and among the first ships fitted with Parsons geared turbines (Brown-Curtis geared turbines in *Furious*), which enabled them to make 31 knots. *Furious*, with a shaft power of 70,096kW (94,000hp), was slightly faster.

First Cruiser Squadron at Heligoland Bight

The failure of the Dardanelles campaign of 1915 ensured that the Russian support operation was abandoned, but construction of the ships continued, though none of them ever entered the Baltic Sea. Instead, *Courageous* and *Glorious* were added to the strength of the Grand Fleet in 1917. On 17 November both were part of the force involved in the skirmish of the Heligoland Bight, the pair with smaller escort vessels forming the First Cruiser Squadron, with *Courageous* as flagship. *Furious* had a different role. During construction the forepart was converted to hold a seaplane

HMS *Courageous* before being converted to an aircraft carrier, the flying-off platform prominent at the front of the ship.

The metamorphoses of HMS *Furious*

1. 1915 – Original profile, altered during construction, with 457mm (18in) single turrets.
2. 1918 – Now with separate fore and aft flight decks, but tower and funnel retained. Restraining net fitted for landing tests.
3. 1942 – Final form with full-length flight deck and vestigial island (from early 1939) for command stations and smoke vents (see picture right).

hangar and launching gear, making it a unique hybrid design. From 8 to 10 aircraft could be carried.

Carrier role

In 1918 take-off platforms (suitable for light aircraft like Sopwith Strutters and Pups, which could get into the air at very low speeds) were mounted on both turrets. *Courageous* was also equipped with minelaying gear, with four sets of mine chutes mounted on the poop deck. An enthusiasm for torpedo armament on large ships, surprising considering how little use had been made of them during the war, saw two pairs of tubes set on the battery deck aft of the funnel, and on *Glorious* a further eight torpedo tubes on the poop deck, making it a kind of giant destroyer. Between November 1917 and March 1918 *Furious* was stripped of its remaining big gun (both ended up as part of the Singapore land defences), and all superstructure aft of the funnel, and an aircraft landing deck was installed, with a hangar below.

After the end of the War, and in the light of the Washington Treaty for fleet reduction, the decision was taken to convert the other two also into aircraft carriers, making use of the long hulls and the powerful machinery. While *Furious* ended up with a complete flat deck and no superstructure, *Courageous* and *Glorious* were given what would become the standard appearance of the aircraft carrier, with deckhouse and funnel set on the extreme starboard to give maximum flight-deck space.

They resembled each other closely, though from 1938 the pole mast on *Courageous* was replaced by a tripod; and its hull extended further beyond the flight deck aft, giving a better arc of fire for AA guns. The original armour protection was not altered, and the flight deck was given 25mm (1in) armour.

Courageous and *Glorious* functioned very effectively as carriers, but *Courageous* was sunk by torpedoes from *U-29* on 17 September 1939, and *Glorious* was sunk by gunfire from *Scharnhorst* and *Gneisenau* on 8 June 1940. *Furious* survived World War II and was scrapped in 1948.

Specification

Dimensions	Length 239.6m (786ft), Beam 24.7m (81ft), Draught 6.78m (22ft 3in), Displacement 17,527 tonnes (19,320 tons); full load 20,829 tonnes (22,960 tons)
Propulsion	18 Yarrow boilers, 4 Parsons geared turbines developing 69,932kW (93,780hp), 4 screws
Armament	4 381mm (15in) guns, 18 102mm (4in) guns, 2 76mm (3in) AA guns, 2 533mm (21in) torpedo tubes (underwater)
Armour	Armoured deck 76–25mm (3–2in), Belt 76–51mm (3–2in), Bulkhead 76–37mm (3–1.5in), Turrets 330–114mm (13–4.5in), Conning tower 254mm (10in)
Range	5920km (3200nm)
Speed	31 knots
Complement	842

Hood (1920)

Of majestic appearance and fine proportions, *Hood* was seen as the 'ultimate' battlecruiser. Completed too late for World War I, by the next war its deficiencies were apparent even before its fatal confrontation with *Bismarck*.

News in 1915 that Germany was building battlecruisers with 381mm (15in) guns – the *Mackensen* class – prompted the British Admiralty to commission designs for a new class of ship that could outperform them.

Before anything was finalised, the Battle of Jutland was fought and the inadequacy of the British battlecruisers in a heavy gun battle was clearly revealed. Admiral Fisher's doctrine that 'Speed is the best protection' was shown to be false and it was evident that the whole concept of the battlecruiser had to be rethought. Really clear thinking might have resulted in dropping this type of ship, but the Admiralty remained keen on a high-speed, heavy-gunned

***Hood* was the last battlecruiser to be built for the Royal Navy. A post-war generation of even larger battlecruisers was planned but cancelled under the terms of the 1921 Washington Treaty.**

Mast
The mast configuration was often altered: the top section of the mainmast was removed by 1941. *Hood* was the last RN battleship to have masthead control tops.

Hull
The long hull made the ship subject to heavy bending stresses. Ratio of hull to beam was 8.239/1.

Aircraft
In 1932 an aircraft crane and telescopic catapult were experimentally mounted on the poop deck, but not retained.

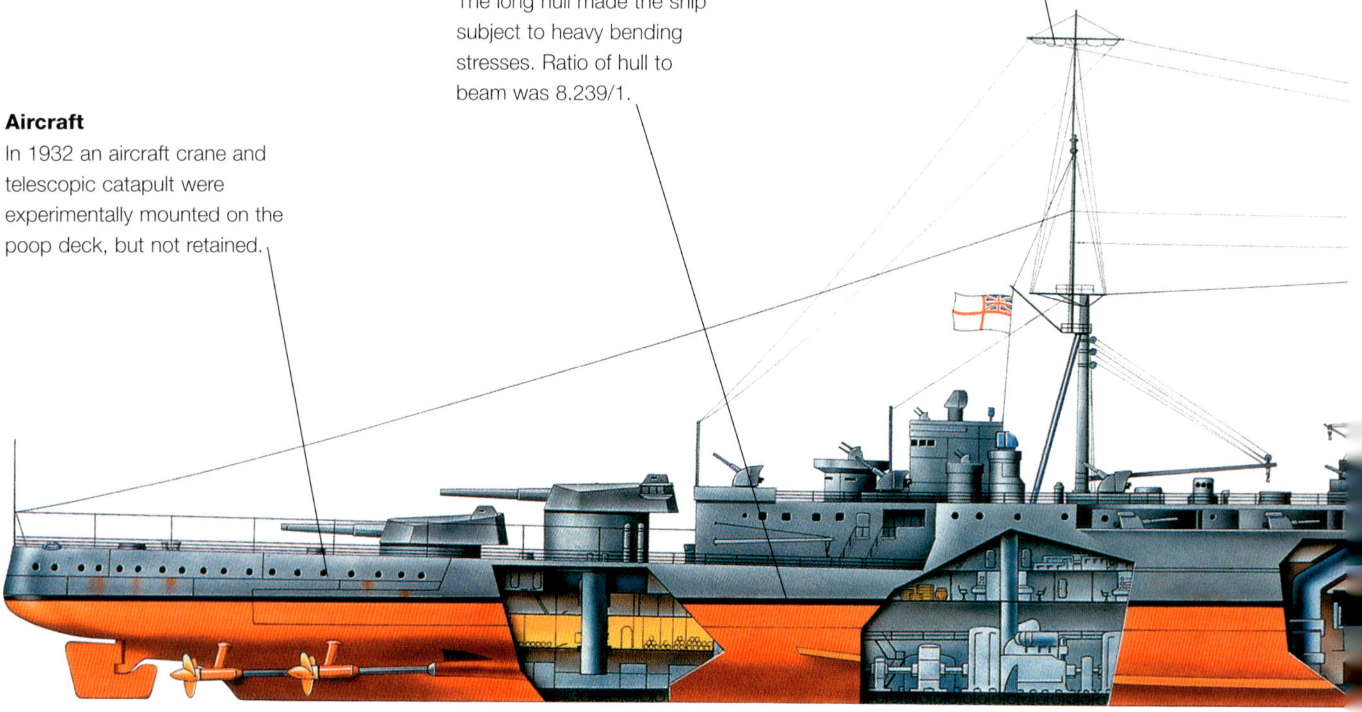

warship, while accepting that it must have greater armour protection than the original design for *Hood* provided.

On redrawn plans, *Hood* was laid down at John Brown's yard, Clydebank, on 1 September 1916 and launched on 22 August 1918. It was to have been the first of four but construction of the others was abandoned when it was learned that the Germans had stopped work on their battlecruisers. *Hood* was completed on 5 March 1920 at a cost of £6,025,000, and commissioned in May.

Formidable weaponry

The eight 381mm (15in) guns were the same as those fitted in the *Queen Elizabeth* class. In total 16 140mm (5.5in) guns were planned for, but only 12 were installed, later cut to 10 in 1939 and six in 1940. AA armament was steadily increased from 16 quadruple-mount 40mm

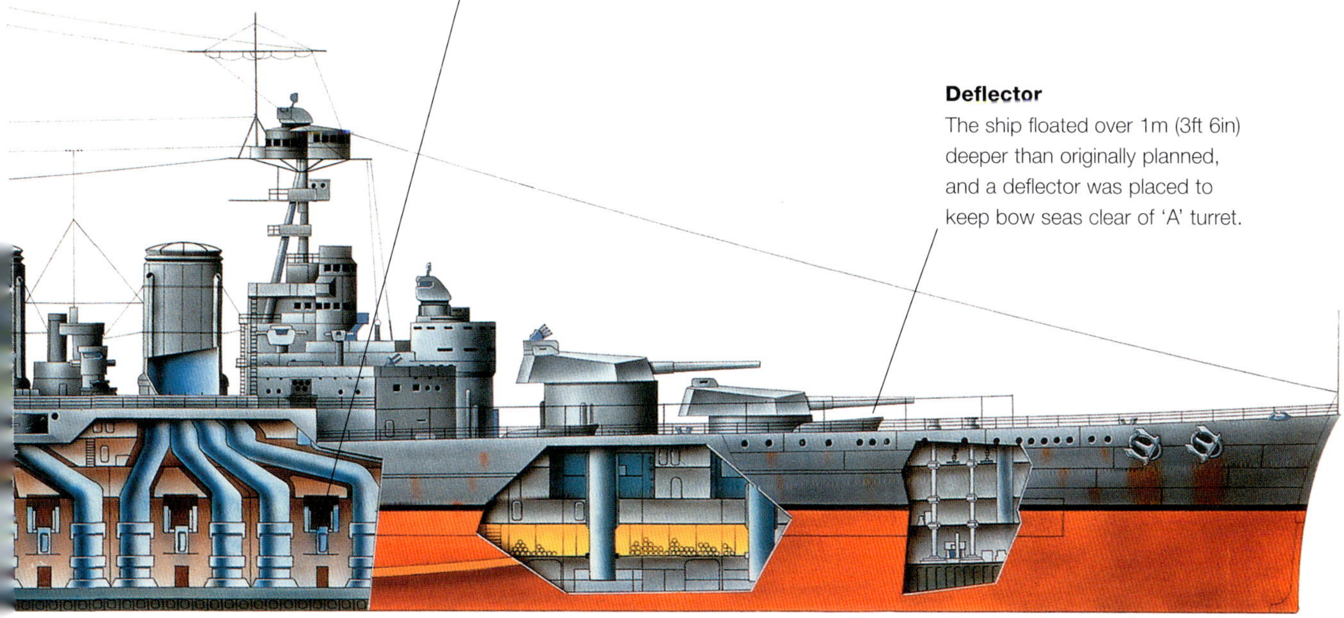

Hood as photographed in 1924 off the coast of Australia. Accompanied by the battlecruiser *Repulse*, *Hood* set out on a world cruise in November 1923. The objective of the cruise was to remind the Dominions of their dependence on British sea power. They returned home 10 months later having visited South Africa, India, Australia, New Zealand and Canada.

(1.6in) guns added in 1931, to 24 from 1939, along with four 102mm (4in) guns (also 1939) increased to eight in 1940. The 15 machine guns installed in 1939 were supplemented by 20 quadruple-mount machine guns in 1940. Two 533mm (21in) underwater torpedo tubes were removed in 1939, but four of the same calibre, mounted on the upper deck abreast of the mainmast, were retained. *Hood* was the first British capital ship to have Yarrow small-tube boilers (24 in all), oil-fired.

Armour belt

In the light of *Hood*'s destruction, the extent and thickness of its armour plating has always generated interest. The revised design incorporated thickening of the belt from 203mm (8in) to 305mm (12in), of the barbettes from 229mm (9in) to 305mm (12in). The main deck was 51–38mm (2–1.5in), with 76mm (3in) over the magazines.

Specification

Dimensions	Length 262m (860ft 7in), Beam 31.8m (104ft 2in), Draught 9.8m (32ft), Displacement 37,376 tonnes (41,200 tons); 41,004 tonnes (45,200 tons) full load
Propulsion	24 Yarrow small-tube boilers, Brown-Curtis geared turbines developing 107,381kW (144,000hp)
Armament	8 381mm (15in) guns, 12 140mm (5.5in) guns, 4 102mm (4in) AA guns, 6 533mm (21in) torpedo tubes
Armour	Belt 305–152mm (12–6in), Bulkheads 127–102mm (5–4in), Deck 76–19mm (3–0.75in), Barbettes 305–127mm (12–5in), Turrets 381–279mm (15–11in), Conning tower 279–229mm (11–9in)
Range	9260km (5000nm) at 18 knots
Speed	31 knots
Complement	1433

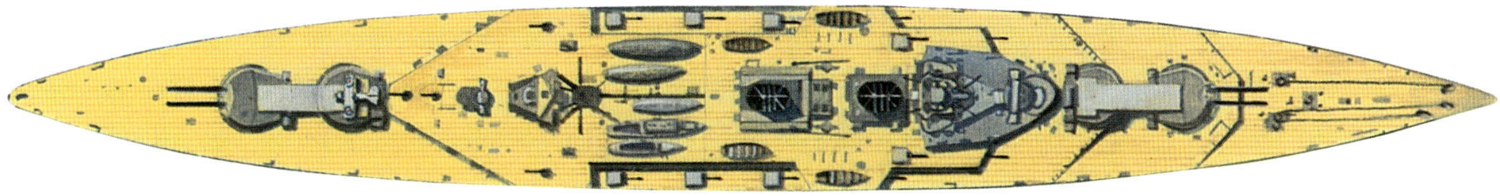

The deck plan shows the original layout of the ship. Note the flying-off platforms for slow-speed Fairey Flycatcher spotter planes, on 'B' and 'X' turrets.

Total weight of armour carried was 12,292 tonnes (13,550 tons), 30.3 per cent of design weight. This compared favourably to the armour weight of British World War I battleships: *Royal Oak*'s armour weight, 7484 tonnes (8250 tons), was 32 per cent of design displacement, but unfavourably to *Bismarck*'s 15,654 tonnes (17,256 tons), making 38.2 per cent of displacement.

In 1939 a complete conversion programme had been considered, including new engines, improved horizontal and underwater protection, a rebuilt bridge and foremast and replacement of the secondary 139mm (5.5in) guns by more modern 132mm (5.2in) QF guns. The expense of this was considered too great and the principal modifications were the removal of the two forward 139mm (5.5in) guns and the installation of 102mm (4in) and 20mm AA guns. By early 1941 the ship also carried Type 279 radar on the mainmast, Types 282 and 285 on the gunnery control equipment and Type 274 on the foremast.

Collision at sea

From 1920–29 *Hood* served with the Home Fleet on exercises and ceremonial duties, including a 10 month world cruise with HMS *Repulse* in 1923. A modernisation programme at HM Dockyard Portsmouth lasted from June 1929 to May 1931, after which it rejoined the Home Fleet. On 23 January 1935 it collided with *Renown* and went to Portsmouth for repairs. After a refit between February and September 1939 it went into war service, suffering slight bomb damage in the North Sea on 23 October the same year.

Operations included the chase of *Scharnhorst* and *Gneisenau* in the English Channel in November 1939, and in June–July 1940 it was flagship of 'Force H' in the Mediterranean, including the attack on the French fleet at Mers-el-Kebir, 3 July. It then returned to the Home Fleet. A further refit was undertaken at Rosyth between January and March 1941. On 22 May 1941 it was flagship of Vice Admiral Holland's Battle Cruiser Force, sent with HMS *Prince of Wales* to intercept *Bismarck* and *Prinz Eugen*, south of Greenland. While exchanging gunfire with the German ships at a range of around 22,860m (25,000yd), *Hood* exploded at 06:00 on the 24th and sank a few minutes later. Only three of the crew survived.

Sinking of the *Hood*

The encounter south of Greenland was the first action between capital ships since Jutland. *Bismarck*'s main guns were 380mm (15in) and *Prinz Eugen*'s were 203mm (8in). *Hood* should have had sufficient staying power to absorb hits and return fire, and its rapid destruction was a cause of intense speculation as well as shock. Two enquiries considered the evidence, *Hood*, with armour of 1914–18 standard, was facing high-velocity shells of 798kg (1759lb) from a better-protected opponent.

Initially *Hood* and *Prince of Wales* opened fire on *Prinz Eugen*, mistaking it for *Bismarck*. *Bismarck* fired five salvoes at *Hood*. A fire had broken out on *Hood*'s boat deck, probably caused by a hit from *Prinz Eugen*, but was not associated with the explosion. The inquiries concluded that a single shell from *Bismarck*'s fifth salvo penetrated either the main deck, 48mm (1.875in) thick, or the upper armoured belt, 127mm (5in) thick, reaching far enough to set fire to the after magazines, which contained around 85.2 tonnes (94 tons) of cordite.

Nagato (1920)

The war of 1905 with Russia had confirmed Japan as a major naval power. With the *Nagato* class it possessed two of the largest, most powerful and fastest battleships in the world.

Japanese naval chiefs studied the Battle of Jutland closely when the *Nagato* class was at the design stage, resulting in some delay while the lessons were absorbed by a team led by Captain Yuzuru Hiraga. *Nagato* was laid down on 28 August 1917 at the Kure Kaigun Kosho naval basin, launched on 9 November 1919 and completed on 25 November 1920.

Nagato was the first ship to be armed with 406mm (16in) quick-firing guns. With L/45 barrel length and a weight of 100 tonnes (111 tons), they elevated (from 1936) to 43 degrees and had a range of 42,000m (45,930yd). The shells weighed 993kg (2189lb) and rate of fire was four rounds in three minutes. The armour plating was on the American principle of maximum

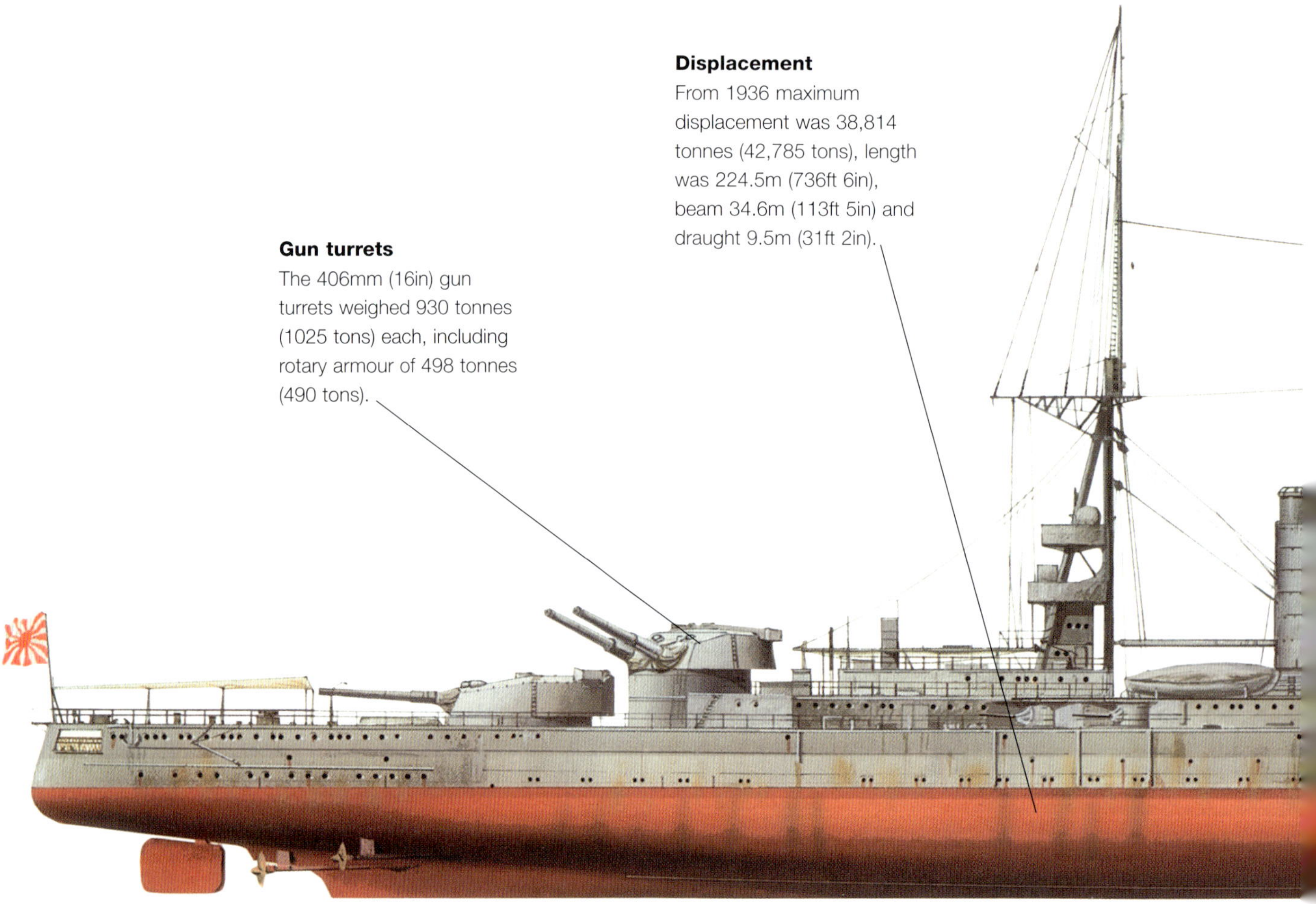

fortification to the vital areas and little or nothing elsewhere, except that the lower deck was well protected with a thickness of 76–71mm (3–2.8in). Among the post-Jutland refinements of design was better anti-flash protection, with longitudinal anti-flash bulkheads between the guns, and in the lower levels of the turrets.

The profile shows *Nagato* between 1925 and 1934; the face-on drawing shows the torpedo bulges applied in the 1934–36 refit, widening the beam from 28.9m (95ft) to 34.6m (113ft 6in). A triple bottom was also installed.

'Pagoda' structure
The tower-foremast structure went through two major changes, resulting in an ever-higher 'pagoda' form that was typical of large Japanese battleships in the 1930s.

Radar
From 1944 *Nagato* carried radar antennae on both masts: Type 1 Model 3 (air search) and Type 2 Model 1 (surface search).

Specification

Dimensions	Length 215.8m (708ft), Beam 29m (95ft 3in), Draught 9.1m (29ft 9in), Displacement 30,159 tonnes (33,245 tons); 34,927 tonnes (38,500 tons) full load
Propulsion	21 boilers (15 oil, 6 coal), 4 Gihon turbines developing 59,656kW (80,000hp), 4 screws
Armament	8 406mm (18in) guns, 20 140mm (5.5in) guns, 4 76mm (3in) AA guns, 8 533mm (21in) torpedo tubes
Armour	Belt 305–102mm (12–4in), Deck upper 44–25mm (1.7–1in); lower 75–50mm (2.9–2.7in), Barbettes 300mm (11.8in), Turrets 356mm (14in), Conning tower forward 371mm (14.6in), aft 97mm (3.8in)
Range	10,550km (5700nm) at 16 knots
Speed	26.75 knots
Complement	1333

Formidable threat

Nagato and its sister ship *Mutsu* (completed October 1921) outgunned any battleship in Western fleets. Even with the completion of USS *Maryland* in July 1921, carrying guns of the same calibre, the Japanese guns fired heavier shells and had a longer range. During 1921 negotiations for the Washington Naval Agreement were in progress to limit the size and number of capital ships, and construction stopped on six American battleships of the original *South Dakota* class. The Agreement allowed the almost-completed *Mutsu* to be commissioned. Of the two other Japanese battleships already laid down, one was stopped and the other eventually became the carrier *Kaga*.

Throughout its career *Nagato* carried its four main gun turrets, but in other respects underwent many changes. These began with a refit in 1924. In original form, the foremast was a multi-level tower mast built around six major struts and rising high above the navigation bridge.

Nagato as it appeared in early 1944. An aircraft catapult was installed between the mainmast and 'X' turret.

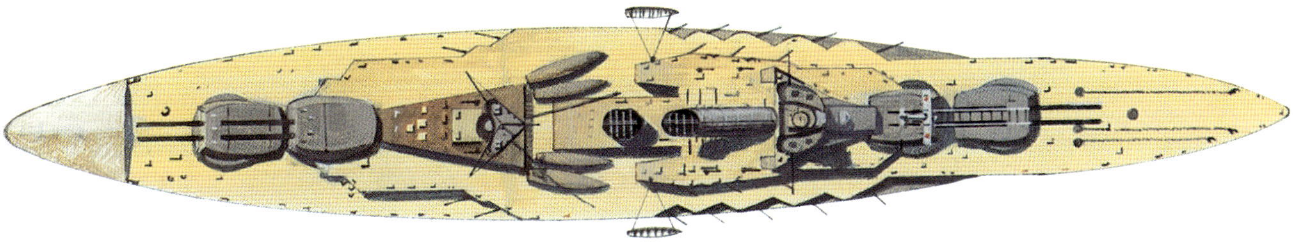

A deck plan of the 1924–34 layout. Note the prototype Heinkel catapult fitted on 'B' turret.

The central stem had a lift inside it, but the arrangement was disturbed by smoke and heat from the raised fore-funnel.

A cowl fitted in 1922 was insufficient and in 1924 the funnel was altered in a bent-back formation to distance it from the tower. A second derrick was fixed to the aftermast to lift and lower floatplanes, and in 1925–26 a German Heinkel catapult was fitted to 'B' turret and became a prototype for Japanese designs.

Armour reinforcement

A much more extensive modernisation was undertaken in 1934–36. Already in 1933 extra AA guns and a catapult abaft the aftermast had been fitted. Now the horizontal armour was reinforced to a maximum 207mm (8.15in), torpedo bulges 2.8m (9ft 2in) wide were applied, new boilers and engines were fitted along with a single funnel set well back from the tower and the hull was lengthened by 8.7m (28ft 6in) at the stern.

The torpedo tubes were taken out and new fire control systems were installed, along with further light AA guns. The main guns had their maximum angle of elevation raised from 30 to 43 degrees. An aircraft crane that could fold to lie on the deck was fitted on the port side of 'X' turret. The original inward-curving bow was replaced by a straight clipper bow (first tried out on *Mutsu*). Searchlight platforms were erected on each side of the funnel.

Changed profile

By 1936 the profile of *Nagato* was very different to that of 1933. Rangefinding equipment was installed, at first on 'B' and 'X' turrets and later above the bridge, along with radar aerials on both masts from 1944. Also in 1944 a high platform was built between the foremast and the funnel as a deck for further AA guns. By this time aerial attack was a deadly menace.

In World War II *Nagato*'s main actions were operations off Midway Island in June 1942, and the Leyte battles of 22–25 October 1944 when it sank the American escort carrier *Gambier Bay* and three destroyers. It returned to the Yokohama naval base and remained at anchor there, with the funnel cap and the after topmast taken down. On 18 July 1945 it was heavily damaged by aircraft strikes. After the war it was used as a target ship in the Bikini Atoll atom bomb tests. The second 'Baker' test on 25 July 1946 left it a wreck and it sank four days later.

Changing times

In the post-*Dreadnought* period from 1907 to 1921, the Japanese Imperial Navy acquired eight battleships and four battlecruisers. Official policy from 1910 was known as 8/8, meaning two capital ship squadrons of eight battleships and eight battlecruisers. Only one of these ships, the battlecruiser *Kongo* (1911), was built outside Japan (in Great Britain), as a deliberate move to acquire a prototype, though many Japanese ideas were incorporated in the design.

By 1919 it was realised that the pre-war ships were not fit for modern battle conditions, and *Nagato* and *Mutsu* were regarded as the first two of a new battleship squadron of eight ships. Though the USA and Japan had been allies in World War I, a trans-Pacific rivalry between these powers was a growing element in global politics. Japanese strategists were exploring the problems of a Pacific war, with its requirements of long range as well as heavy firepower, and battleships were an important element.

Rodney (1927)

Of unusual design with all the heavy guns mounted forward, *Rodney* was one of the two Royal Navy battleships built under the terms of the Washington Naval Treaty of February 1922, and was the only British warships to carry 406mm (16in) guns.

The terms of the Washington Treaty allowed Great Britain to build two new battleships, of a displacement no greater than 31,751 tonnes (35,000 tons). Both were laid down on 28 December 1922 – HMS *Nelson* at Armstrongs on the Tyne and *Rodney* at Cammell Laird, Birkenhead. *Rodney* was launched on 17 December 1925, and completed in August 1927 at a cost of £7,617,799.

Machinery
Rodney's machinery arrangement was unusual, with the eight Admiralty boilers placed abaft the engine room with its two Brown-Curtis geared turbines. This was in order to keep the funnel as far away from the tower as possible.

AA guns
The final AA armament fitted was 16 40mm (1.6in), 48 2-pounder pom-pom and 61 20mm (0.79in) guns.

Design ingenuity

Much thought and ingenuity went into the design in order to achieve the best balance of gunpower and protection. Lightweight 'D' steel was used in hull construction, and Douglas fir rather than teak was used for decking, with aluminium and fireproofed plywood for interior fitments. The four battlecruisers proposed for 1921, and cancelled, would have had Mark I triple-mounted 406mm (16in) guns, and these were incorporated in the new battleships along with some other material that had been prepared. With a barrel length of L/45, they weighed 117 tonnes (129 tons) and had a range of 41,600m (45,490yd) at a maximum elevation of 44 degrees. The shells weighed 907kg (2000lb) and two rounds a minute could be fired. The guns were fitted in single cradles and could be individually aimed.

Secondary armament was 12 152mm (6in) guns, no longer in casemates but in individual turrets, and the first in the RN to be power-worked; six 119mm (4.7in) AA guns and eight 2-pounder pom-poms. Two 622mm (24.5in) torpedo tubes were fitted, the only British use of this calibre. The five-deck tower design, with integrated mast, had also been planned for the battlecruisers and carried directors for the various gun groups, Admiral's bridge, torpedo control station, signalling and navigation bridges. A pedestal above the bridge turret carried an AA gun position initially, but in 1935–36 the platform was enlarged to carry new fire-control equipment. *Rodney* and *Nelson* were the last British battleships to be built with a separate conning tower. Innovations in provision included oil-fired galleys and electric ovens (rather than coal) and the standard of crew

The camouflage scheme was intended to reduce the ship's visibility in Atlantic waters. *Rodney*'s torpedo rooms were located forward of 'A' turret. It was one of very few battleships to strike another with a torpedo, hitting *Bismarck* during the battle on 27 May 1941.

Specification

Dimensions	Length 216.8m (710ft), Beam 32.4m (106ft), Draught 9.1m (30ft), Displacement 30,799 tonnes (33,950 tons); 34,473 tonnes (38,000 tons) full load
Propulsion	8 3-drum boilers with superheaters, Brown-Curtis geared turbines developing 33,556kW (45,000hp), 2 screws
Armament	9 406mm (16in) guns, 12 152mm (6in) guns, 6 119mm (4.7in) AA guns, 8 2-pounder pom-poms, 2 622mm (24.5in) torpedo tubes, underwater
Armour	Belt 356mm (14in), Bulkheads 356–76mm (14–3in), Barbettes 381mm (15in), Turret faces 406mm (16in), Deck over magazines 158mm (6.25in); over machinery 76mm (3in)
Range	26,500km (14,300nm) at 12 knots
Speed	23 knots
Complement	1314

accommodation, including recreation and drying rooms, was far above that of any previous RN ship.

'Ugly but impressive'

With these ships the Royal Navy opted for the American 'all or nothing' armour system. The horizontal armour above the engine room was 76mm (3in) thick (it was pierced by a bomb in 1940, which failed to explode) and above the 406mm (16in) magazines it was 159mm (6.25in). For the first time, side armour was not fitted as an external belt, but on the inner side of the steel hull, angling inwards from the armoured deck: the effect of this was intended to reduce the penetrative power of an incoming shell and also to deflect plunging shells. Torpedo bulges were not applied.

Steam was supplied by eight small-tube boilers, six of 4176kW (5600hp) and two of 3803kW (5100hp). Speed at 42.6km/h (23 knots) was not exceptional. Described as 'ugly but impressive', *Rodney* was a difficult ship to handle, notoriously slow in turning and requiring a

Shells for the 406mm (16in) guns being loaded from a supply vessel during World War II. Each one weighed 1198kg (2641lb). Details of rangefinding and radar equipment can be seen.

Guns on the *Incomparable* class

Much of the design of the four *Incomparable* class battlecruisers laid down in 1921 (such as the *Nelson*, pictured), but subsequently cancelled, was incorporated in the plans for *Rodney*, though it displaced around 10,886 tonnes (12,000 tons) less. These had also been intended to carry nine 406mm (16in) guns, though the third turret was to be between the tower and the two funnels, with the fore and after superstructure angled to provide it with a wide arc of fire. Their engines would have been much more powerful and their speed some 22km/h (12 knots) higher. All *Rodney*'s main guns were sited forward of the tower in order to concentrate the magazine armour in as limited a space as possible. In theory the guns could fire abaft the beam but in practice this proved impossible, the blast smashing even reinforced glass windows and damaging the tower structure. The 40 degree elevation was used on occasion to fire 1198kg (2641lb) shells at distant approaching aircraft up to 32,000m (35,000yd) away.

considerable degree of opposite helm in order to maintain a straight course.

Service

Rodney was based with the Home and Atlantic Fleets from 1927 to 1939. Its war service was intensive, based first at Scapa Flow, then Rosyth from 13 September 1940. From November 1940 it was on Atlantic convoy escort while on its way to the USA for an engine repair when it was detached to join in the hunt for *Bismarck* in May 1941. With *King George V*, *Rodney* opened fire on the German battleship at 08:07 on 27 May and participated in its destruction. It then resumed the voyage to Boston where engine repairs were carried out during June–September 1941.

After repair *Rodney* served briefly in the Mediterranean before returning to Britain in November. A complete refit was undertaken between December 1941 and May 1942, then it returned to the Mediterranean to support the North African landings in November.

With *Nelson* it remained in the Mediterranean until October 1943, when it resumed service with the Home Fleet, bombarding shore targets in France and the Channel Islands, and escorting convoys to Murmansk. *Rodney* steamed 251,000km (156,000 miles) on war service. From December 1944, with engines badly needing an overhaul, it remained at Scapa Flow then passed into reserve after the end of the war. In February 1948 it was sold for breaking.

Deutschland/Lützow (1933)

Signalling the re-creation of a sea-going German Navy, *Deutschland* (renamed *Lützow* in 1939) was the first of a class of 'pocket battleships' that included *Admiral Graf Spee,* and was the first major warship to be powered by diesel engines.

After World War I Germany was left with eight obsolete battleships and a limit, imposed by the Treaty of Versailles, of 9072 tonnes (10,000 tons) on new construction, which could only begin after 1924. Early in the 1920s work began on establishing whether an effective heavy-gun ship could be constructed within this limit. At first it proved impossible to provide the necessary weight of armament and the weight of armour that would give staying-power in combat.

In 1927, the concept of a fast *Panzerschiff* (armoured ship) began to crystallise. It would have greater speed than a battleship, with thicker armour and heavier guns than a cruiser. Guns of 280mm (11in) calibre were envisaged, and high speed was to come from diesel engines. With a full specification drawn up, *Deutschland* was laid down at the Deutsche Werke, Kiel, on 5 February 1929, launched on 19 March 1931 and completed on 1 April 1933. Although officially recorded as 9072 tonnes (10,000 tons), the ship's actual displacement was 10,614 tonnes (11,700 tons).

Weight-saving measures

Every effort was made to save weight, helped by technical improvements. Electric welding was used in building the

Mainmast
From a 1935 refit the mainmast was rigidly connected to the funnel.

Armour covers
Protective armoured covers were fitted over the torpedo tubes shortly after commissioning.

Above: *Graf Spee* as it was in 1939, showing various differences in detail: funnel, aircraft mounting, radar, AA guns, etc., to *Deutschland/Lützow.*

Below: *Lützow* in 'splinter' camouflage paint, 1942, when it was based in the fjords of northern Norway. The FuMO2 radar frame was fitted to the top of the tower in that year.

Aircraft
Deutschland was the only ship in the class to carry its two aircraft between tower and funnel (fitted in 1935).

hull, resulting in a hull weight of 3557 tonnes (3921 tons), 15 per cent less than if riveting had been used. The main guns, in two triple turrets, had a barrel length of L/54.5 and weighed 47.5 tonnes (52.4 tons), firing a shell of 315kg (694lb). Their range, 42,600m (46,588yd), exceeded that of any other naval gun at the time – a major tactical advantage. Secondary armament consisted of eight 152mm (6in) quick-firing guns in laterally-placed single-mount shields on the forecastle deck (which extended four-fifths of the ship's length) and three 88mm (3.5in) AA guns. Eight 500mm (19.7in) torpedo tubes were fitted in two sets of four on the upper deck, with armoured covers; later 533mm (21in) tubes were fitted.

The term 'pocket battleship', first used outside Germany, was not intended as a compliment but it became clear that *Deutschland* and its two sister-ships, *Admiral Scheer* (1934) and *Admiral Graf Spee* (1936), had to be taken seriously as commerce raiders. Apart from providing a maximum speed of 52km/h (28 knots), the engines gave a wide operational range of 52,782km (28,500nm) at 18.5km/h (10 knots). The power came from four MAN nine-cylinder double-acting two-stroke motors, via two Vulcan reduction gear-boxes.

Few modifications were made before 1935, by which time the Third Reich's military ambitions were apparent. In 1934 the port side derrick was replaced by a lattice crane. In 1935 a mainmast was attached abaft the funnel, AA gun control was fitted, an aircraft catapult installed between the bridge tower and funnel, and twin-mount 88mm (3.5in) AA guns replaced the original singles. Further changes in

Specification

Dimensions	Length 186m (610ft 3in), Beam 20.6m (67ft 6in), Draught 7.2m (23ft 7in), Displacement 10,659 tonnes (11,750 tons); 14,424 tonnes (15,900 tons) full load
Propulsion	4 9-cylinder double-acting 2-stroke MAN diesel engines developing 40,268kW (54,000hp), Vulcan gearboxes, 2 screws
Armament	6 280mm (11in) guns, 8 150mm (5.9in) guns, 3 88mm (3.5in) guns, 8 500mm (19.7in) torpedo tubes
Armour	Belt 80–60mm (3.1–2.4in), Bulkheads 45–40mm (1.8–1.6in), Deck 45–40mm (1.8–1.6in), Barbettes 150–50mm (5.9–2in), Turrets 140–85mm (5.5–3.3in), Conning tower 150–50mm (5.9–2in)
Range	16,120km (8700nm) at 19 knots
Speed	28 knots
Complement	619

A trio of battleships

For a time the trio of *Deutschland*, *Admiral Scheer* and *Admiral Graf Spee* occupied a unique position – more powerful than any cruiser, faster than any battleship and with an unmatched range of operation. Only the older British battlecruisers *Hood*, *Repulse* and *Renown* exceeded them in both speed and firepower. Nevertheless, it was political considerations rather than strategic planning that dictated their design and construction. Six of the class were planned in November 1932, and assembly of materials began for the fourth and fifth, but were used instead for the full-size battleships *Scharnhorst* and *Gneisenau*. *Admiral Graf Spee* took refuge in Montevideo, Uruguay, after the Battle of the River Plate, 13 December 1939, and was destroyed by its own crew on the 17th (see picture).

Admiral Scheer had an active career through World War II before succumbing to an air attack by RAF bombers while undergoing a refit at Kiel on 9 April 1945. After five heavy hits, it capsized and sank.

Admiral Graf Spee in 1939. It and *Admiral Scheer* had angular bridge tower structures, different to *Lützow*'s. Seetakt radar, primarily for gunnery, was fitted to *Graf Spee* in 1939: the range-finder or telemeter is prominent at the top.

1937–38 included the fitting of new cranes on each side, the addition of a slightly tilted cap to the funnel (considerably heightened at the fore-end in 1942) and the placing of large searchlights on a platform mounted round the funnel. In its final form the bridge tower, tapering towards the top, was surmounted by a wide range-finder which was in turn topped by a wireless mast.

Sustained damage

On commissioning, *Deutschland* became flagship of the German Navy until 1936. In 1936–37 it made patrols off Spain after international maritime controls had been agreed and took two hits from Spanish Republican aircraft on 29 May 1937, causing considerable damage and 31 deaths. From 29 September 1939 it was engaged in commerce raiding in the Atlantic. Its name was changed to *Lützow* on 15 November 1939. On 11 April 1940 the British submarine *Spearfish* hit it with a torpedo off Norway. With a severely damaged stern and no propellers or steering it was towed back to Kiel. On 13 July 1941, attempting an entry to the Atlantic, it was again badly damaged by an aerial torpedo from an RAF Beaufort, and returned to Kiel for repairs until January 1942.

Through 1942 and until 29 September 1943 it was based in North Norway, with a return to Kiel for repairs from 9 August to 9 November 1942 following grounding in Tjeldsund. In December 1942 it was involved in the Battle of the Barents Sea against the British cruisers *Sheffield* and *Jamaica*. From October 1943 it remained in the Baltic Sea, based at Gdynia, protecting the retreat of German troops and shelling Russian inland positions. A bomb attack on 15 April 1945 left the ship disabled while rearming. Scuttled by its crew on 4 May, it was eventually destroyed by Russian bomb practice and then scrapped.

Deck plan: the 150mm (5.9in) guns are all on the same level as the flush deck. The compact superstructure offered little room for higher gun emplacements.

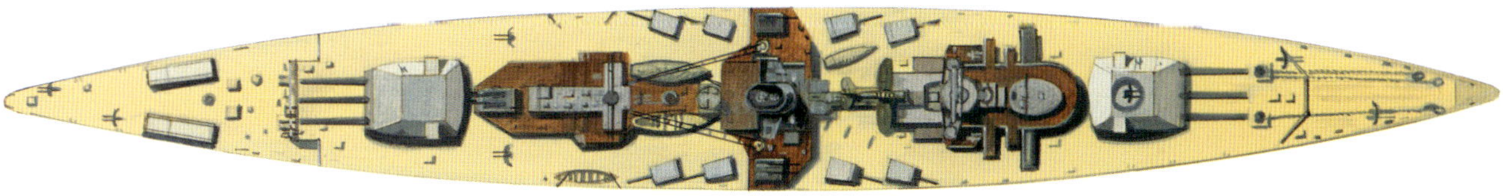

Gneisenau (1938)

Gneisenau and _Scharnhorst_ had been paired armoured cruisers (1906), often in action together during World War I. The tradition was maintained by these battleships of the Third Reich.

In the initial years of the Hitler regime from 1933, German naval plans stuck to the 9072 tonne (10,000 ton) *Panzerschiff* concept. But the French were constructing the notionally 24,040 tonne (26,500 ton) *Dunkerque* – actually well in excess off 26,785 tonnes (29,526 tons) – and the Italians were also planning new battleship construction. In 1934 and 1935 assemblage of materials had begun for construction of pocket battleships 'D' and 'E' at Wilhelmshaven and Kiel respectively. An increase in displacement to 17,236 tonnes (19,000 tons) for these had been secretly agreed in order to provide better armour protection. The displacement was further increased to

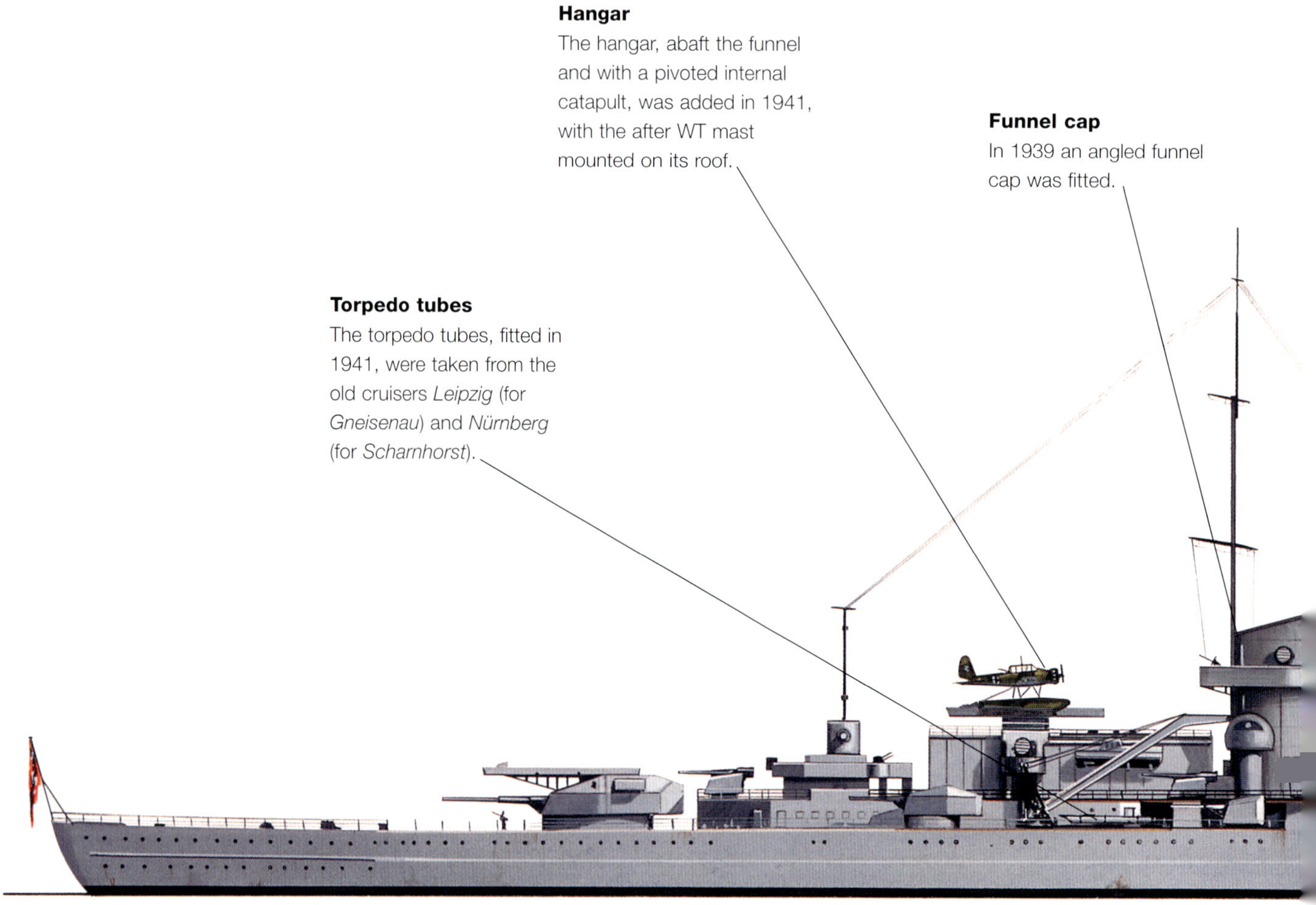

23,587 tonnes (26,000 tons) and the new ships were to carry nine 280mm (11in) guns. Despite this they were still officially designated as *Panzerschiffe*. The controversial Naval Agreement between Britain and Germany of 18 June 1935, allowing the tonnage of the *Kriegsmarine* to be 35 per cent of that of the Royal Navy, legitimised the plan.

Main guns

Gneisenau was laid down at the Deutsche Werke, Kiel, in March 1935, launched on 8 December 1936 and completed on 21 May 1938. Its cost was 146,174,000 Reichsmarks. The main guns were the same as those fitted to the *Deutschland* class, in three triple turrets, though the rotary mountings were of Type C/34 rather than Type C/28 as in the earlier ships. The 12 150mm (5.9in) guns were also as fitted on the *Deutschland*, though eight were in double-mounted turrets and the others in single mount shields. In total 14 105mm (4.1in) double-mount AA guns, 16 37mm (1.4in) double-mount AA guns and eight (increased to 10 in late 1939) 20mm (0.79in) AA guns were also fitted.

Gneisenau was equipped to carry aircraft from the start, with two catapults (reduced to one in 1940) and a capacity of four floatplanes. The initial hull design incorporated a vertical-stemmed bow and double hawse-holes, but in

Gneisenau and _Scharnhorst_ were of very similar appearance, the two differing chiefly in the placing of the mainmast, which in _Gneisenau_, as in _Deutschland,_ was attached abaft the funnel.

Gneisenau in the Korsfjord, close to Bergen, Norway, on 30 December 1940. The ship is on the point of leaving for Gotenhafen (Gdansk) on the Baltic coast.

ocean-going tests the foredeck was frequently awash and 'A' turret put out of use. In the winter of 1938–39 the bow was rebuilt in 'Atlantic' form, higher and with a sharp rake, adding 5.1m (16ft 8in) to overall length. In this form the ship's seagoing qualities were only slightly improved, and further alterations were made in 1942. The hawse-holes, retained at first, were later plated over and replaced by deck hawse-holes. A fore-cap was fitted to the funnel to keep smoke away from the mast.

Deck plan: symmetry is usually noticeable in deck plans, and what is most striking about *Gneisenau*'s arrangement is the way in which the main and secondary guns are given the widest possible arcs of fire.

Official classification

In 1941 *Gneisenau* had 12 20mm (0.79in) four-barrelled guns added to its AA defences, and a large hangar with internal catapult was fitted. Six 533mm (21in) torpedo tubes were installed in two sets of three on the upper deck.

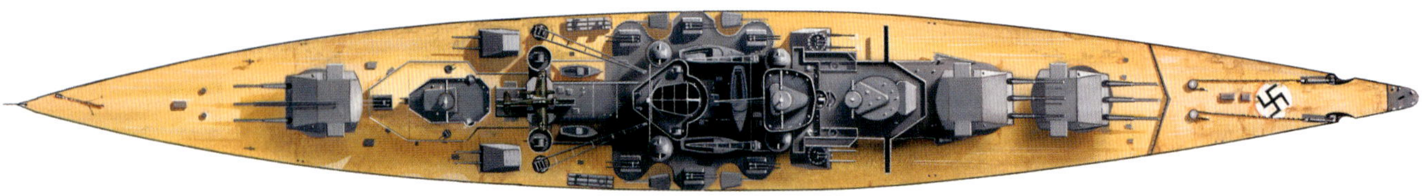

Though sometimes referred to as battlecruisers, *Gneisenau* and *Scharnhorst* were finally classed as *Schlachtschiffe*, or battleships, despite the relatively light 280mm (11in) main guns. In 1942 plans were drawn up to refit *Gneisenau* with nine 380mm (15in) guns in twin turrets, but this was not put into effect. Diesel engines had been planned, but in order to achieve 55.5km/h-plus (30 knot) speeds, high-pressure steam turbines were installed.

Most of *Gneisenau*'s war activity was carried on in conjunction with *Scharnhorst*. On 23 November 1939 they sank the armed merchant cruiser *Rawalpindi* north-west of the Faroes. In supporting the landings at Narvik, Norway, on 9 April 1940, *Gneisenau* exchanged fire with HMS *Renown* and the rear turret was disabled. Repairs had just been completed when the ship struck a mine and had to return to Kiel for further repairs. In June the two German battleships sank the British carrier *Glorious* and the destroyers *Acasta* and *Ardent* off Narvik. While returning to Germany *Gneisenau* was seriously damaged in the bows by a torpedo from the British submarine *Clyde*.

Channel dash

Between 22 January and 23 March 1941 *Scharnhorst* and *Gneisenau* sank 22 merchant vessels on a commerce-raiding cruise in the Atlantic. At Brest on 6 April 1941 it was put out of action for some time by an aerial torpedo hit. After remaining for almost a year at Brest they made the famous Operation Cerberus or 'Channel dash' on 11–13 February 1942, along with the heavy cruiser *Prinz Eugen* and numerous destroyers and E-boats as escorts. *Gneisenau* struck a mine but received only minor damage. On 26 February it was heavily damaged in a bombing raid while dry-docked. Moved to Gotenhafen (Gdansk), it was to be repaired and regunned, and in January 1943 the 280mm (11in) guns were removed for coastal defences. Work ceased in February 1943. On 27 March 1945 it was scuttled by the *Wehrmacht* as a blockship, and from 1951 it was salvaged and broken up.

Specification

Dimensions	Length 229.8m (753ft 11in), Beam 30m (90ft 5in), Draught 8.23m (27ft), Displacement 31,606 tonnes (34,840 tons); 35,289 tonnes (38,900 tons) full load
Propulsion	12 Wagner HP boilers, 3 Germania geared turbines developing 123,040kW (165,00hp), 3 screws
Armament	9 280mm (11in) guns, 12 150mm (5.9in) guns, 14 105mm (4.1in) guns, 16 37mm (1.4in) and 8 20mm (0.79in) AA guns
Armour	Belt 350–200mm (13.8–7.9in), Bulkheads 200–150mm (7.9–5.9in), Deck 50–20mm (2–0.8in), Barbettes 350–200mm (13.8–7.9in), Turrets 350–200mm (13.8–7.9in), Conning tower 350–100mm (13.8–3.9in)
Range	16,297km (8800nm) at 19 knots
Speed	32 knots
Complement	1840

Atlantic raiding

Hitler saw the German Navy's new battleships as 'special monuments of the German war state' and so of great political and prestige value. While his naval commanders, Raeder and Fricke, wanted *Scharnhorst* and *Gneisenau* to carry out further Atlantic raids from Brest once *Gneisenau*'s repairs had been completed, Hitler insisted that the ships should be transferred to more secure bases in Norway. If that were not possible, then they should be taken out of service and their guns transferred to shore bases. This led to Operation Cerberus (see p.182). But the failure of *Lützow* and the heavy cruiser *Admiral Hipper* to secure a victory over the lighter-gunned HMS *Jamaica* and *Sheffield* in the Battle of the Barents Sea on 31 December 1942 caused Hitler to order cessation of work on capital ships. Of *Gneisenau*'s dismounted guns, turret 'Cäsar' is still preserved at the Austrått coastal fort in Norway, where it had been set up as part of the 'Atlantic Wall'.

Scharnhorst (1939)

Often referred to as the first German battleship to be constructed since World War I, *Scharnhorst,* the first of the two to be launched, was actually laid down and commissioned after *Gneisenau.* Though sister-ships, there were numerous minor differences.

Scharnhorst was laid down at the Kriegsmarinewerft, Wilhelmshaven, on 16 May 1935, launched on 3 October 1936 and completed on 7 January 1939. Its cost was 143,471,000 Reichsmarks. Almost immediately in July–August 1939 its bow was rebuilt to 'Atlantic' form. Armament, armour and machinery were essentially the same as *Gneisenau*'s, though *Scharnhorst* had Brown-Boveri rather than Germania turbines. The three propellers were triple-bladed, with a diameter of 4.45m (14ft 6in) and *Scharnhorst*'s speed of 58.6km/h (31.64 knots) gave it a distinct edge over the British *King George V* class, whose maximum was 54km/h (29.2 knots).

The hull was built of ST 52 steel, and a new steel armour was introduced, known as type Wh, with a tensile strength

As commerce raiders, *Scharnhorst* and *Gneisenau* showed deadly effectiveness, sinking 22 merchant ships between 22 January and 23 March 1941. On three occasions British battleships were sighted but avoided. *Scharnhorst* covered an average of 548km (296nm) a day.

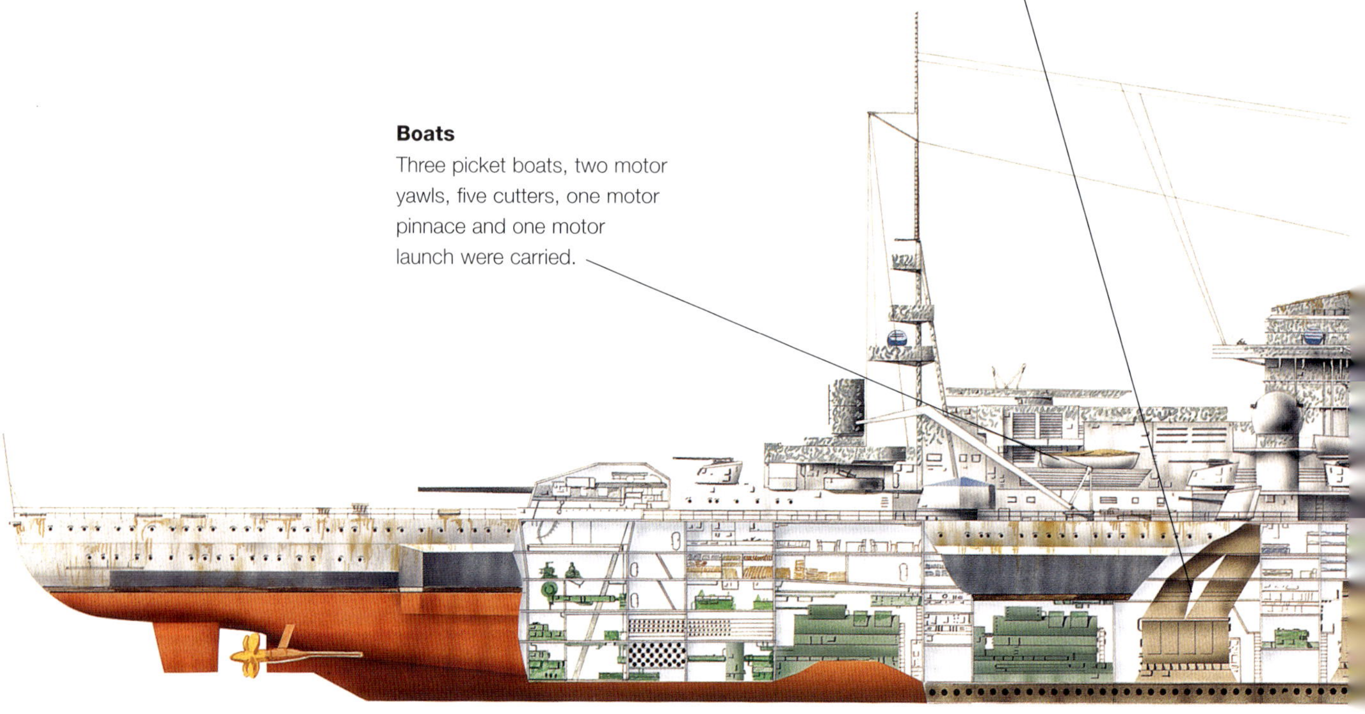

of 85–95 kg/mm², and Ww (65–75 kg/mm²). Apart from their resistant qualities, these steels could be welded, with important advantages in construction speed and weight-saving. Nevertheless, the two ships would ultimately displace over 27,215 tonnes (30,000 tons).

Armour plating

Belt armour had a maximum thickness of 350mm (13.8in), which might be expected to hold against a 356mm (14in) shell, and the armoured deck was 105–20mm (4–0.79in) thick. The armour weighed considerably more than the actual hull at 12,706 tonnes (14,006 tons) compared to 7469 tonnes (8233 tons). But both ships would survive heavy attacks from shells, torpedoes and bombs.

From 1941 *Scharnhorst* was fitted with 12 quadruple-mounted 20mm (0.79in) AA guns. From early 1939 *Gneisenau* had a mainmast attached abaft the funnel, and *Scharnhorst* originally had a taller mainmast in the same

Outline
A general resemblance of outline between *Scharnhorst, Gneisenau, Bismarck* and *Tirpitz* was deliberately planned in order to confuse their identities.

Fuel capacity
Maximum fuel capacity was 5624 tonnes (6200) tons of oil, giving a range of 16,298km (8800nm) at 19 knots.

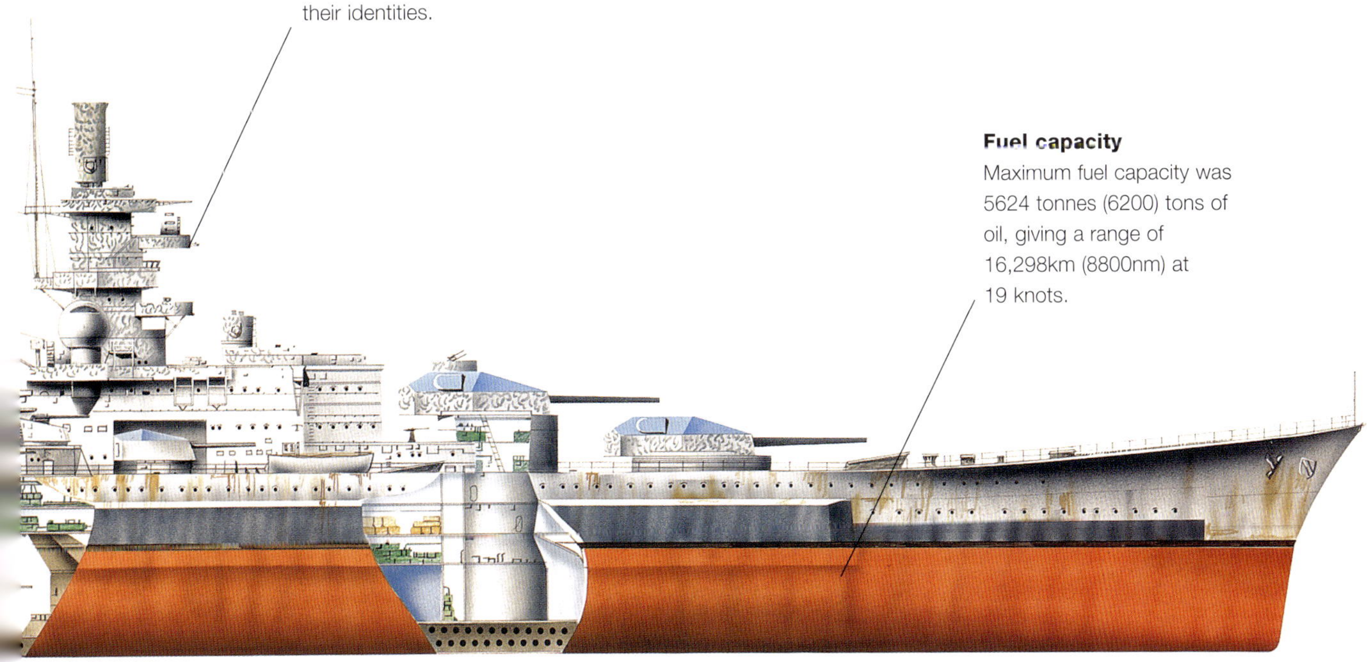

Operation Cerberus

Operation Cerberus was carried out at Hitler's order. Preparations included minesweeping operations along a deep water route to ensure high-speed movement. Six destroyers, 14 T-boats and three motor torpedo boat flotillas were assembled, and the Luftwaffe deployed 250 fighter planes. The intention was known to British naval intelligence but the German squadron was expected to pass through the Straits of Dover by night. Its daytime passage on 12 February, at 50km/h (27 knots) took the British by surprise, and their preparations for attack were insufficient and effectively repelled by the German ships and planes. The mines hit by *Scharnhorst* and *Gneisenau* had been laid by aircraft along the anticipated route. The operation was an undoubted success for Germany, but the strategic background was one of retreat and withdrawal. Hitler was not prepared to risk the capital ships in the Atlantic, and for the rest of the war their role was a limited one.

position but free-standing, though with the searchlight platform built out round it from the funnel casing. In the July–August 1939 alterations, this was changed to a tripod and moved sternwards to a position just forward of the aft control tower.

Specification

Dimensions	Length 235m (772ft), Beam 30m (98ft 5in), Draught 9.7m (31ft 9in), Displacement 29,121 tonnes (32,100 tons), 34,564 tonnes (38,100 tons) full load
Propulsion	12 Wagner HP boilers, 3 Brown-Boveri geared turbines developing 119,312kW (160,000hp)
Armament	9 280mm (11in) guns, 12 150mm (5.9in) guns, 14 105mm (4.1in) guns, 16 37mm (1.5in) and 10 20mm (0.79in) AA guns, 6 533mm (21in) torpedo tubes
Armour	Belt 350–200mm (13.8–7.9in), Bulkheads 200–150mm (7.9–5.9in), Deck 50–20mm (2–0.79in), Barbettes 350–200mm (13.8–7.9in), Turrets 350–200mm (13.8–7.9in), Conning tower 350–100mm (13.8–3.9in)
Range	18,500km (10,000nm) at 17 knots
Speed	31 knots
Complement	1968

Searchlight and radar platforms were mounted, making the clearest difference between the two ships. Both had pole foremasts attached to the main tower. *Scharnhorst* also received an inclined funnel cap.

Aircraft

Originally a second catapult had been installed on the after turret in both ships, with a folding crane mounted on the port side, but this was removed before the end of 1939. At the same time a boom-mounted canvas landing device that could be swung out from the port side was removed. The aircraft carried were three Arado Ar196A-3 (*Gneisenau* for a short time had Heinkel He-114 planes). Range-finding and radar equipment was updated through 1940–42, and by 1942 it had FuMO 27, FuMB1, FuMB 4 SAMOS, FuMB7 and FuMB3 systems installed.

Germany had been working on radar development from 1934 but the systems available in 1943 were less sophisticated than those brought into use by the British and Americans. Five searchlights, triaxially stabilised and operated by a central control station, were fitted.

Naval engagements

In the operations off Narvik in April–June 1940, *Scharnhorst* was first to open fire on the carrier HMS *Glorious*, which was sunk. A torpedo from HMS *Acasta* made a 12 by 4m (38ft by 13ft) gash in the battleship's side and around 2268 tonnes (2500 tons) of water were shipped. At slow speed it made for Trondheim for emergency repairs. Air attacks on it during 10–13 June were a failure, and *Scharnhorst* was able to return to Kiel for full repair. It remained at Kiel until

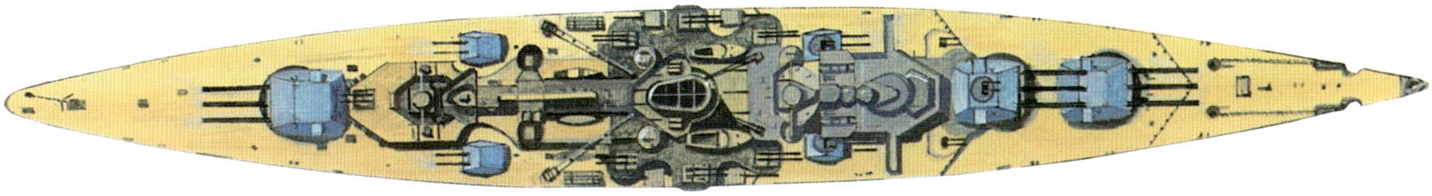

Deck plan: the 150mm (5.9in) secondary guns are clearly visible from this perspective, as are the many smaller 37mm (1.5in) and 20mm (0.79in) AA guns.

22 January 1941, then was based at Brest for the commerce raids of January–March which (with *Gneisenau*) accounted for over 95,254 gross tonnes (105,000 gross tons) of Allied merchant shipping.

In the 'Channel dash' from Brest to Germany between 11 and 13 February *Scharnhorst* struck two mines, the first of which immobilised the ship for 17 minutes, but it made Wilhelmshaven under its own steam. Repaired again at Kiel, *Scharnhorst* joined with *Tirpitz* in bombarding Spitzbergen on 6–8 September 1943. By the end of that year it was the Reich's only operational capital ship. With a force of five destroyers under Rear Admiral Bey, it was sent on 25 December to attack Convoy JW55B en route for Murmansk. In Arctic waters north of North Cape, after an inconclusive encounter with the British cruisers *Norfolk*, *Belfast* and *Sheffield*, *Scharnhorst* came under fire from the radar-directed 356mm (14in) guns of the battleship *Duke of York*, with four escorting destroyers. A combination of torpedoes and shell-fire at increasingly close range finally destroyed *Scharnhorst*, which continued firing from the one surviving turret to its last moments. Its bow was blown off, and it sank at 19:44 on 26 December, in 290m (950ft) of water, with the loss of 1803 men.

Scharnhorst's crew salutes that of *U-47* as it returns to Kiel after sinking HMS *Royal Oak* in September 1939.

Bismarck (1940)

Bismarck and _Tirpitz_ represented the peak of German battleship design – in terms of size, armament, speed and staying-power. But _Bismarck_'s first combat mission, though including a major success, was also its last.

While Adolf Hitler contemplated a war against Great Britain, he told his naval chiefs that it would not be until 1944–45, and when war actually broke out in September 1939 the *Kriegsmarine* had only two modern battleships plus four 'pocket battleships'. But two very large warships had been launched, and two even larger ones had just been laid down (though their construction was abandoned at the end of 1942).

Bismarck was laid down at Blohm & Voss, Hamburg, on 1 July 1936, launched on 14 February 1939 and completed on 24 August 1940. Its cost was 196,800,000 Reichsmarks. The chief designer, Dr Hermann Burkhardt, who was also responsible for *Gneisenau* and *Scharnhorst*, was no longer operating within constraints other than the essential considerations of balancing firepower, speed, protection and propulsive power. The main guns were eight 380mm (15in) mounted in twin turrets, arranged to keep the guns as far apart as possible to avoid shocks and interferences.

Barrel length was L/47 and weight was 97.5 tonnes (107.5 tons). Range was 36,200m (39,589yd) at an elevation of 35 degrees (giving a low trajectory and

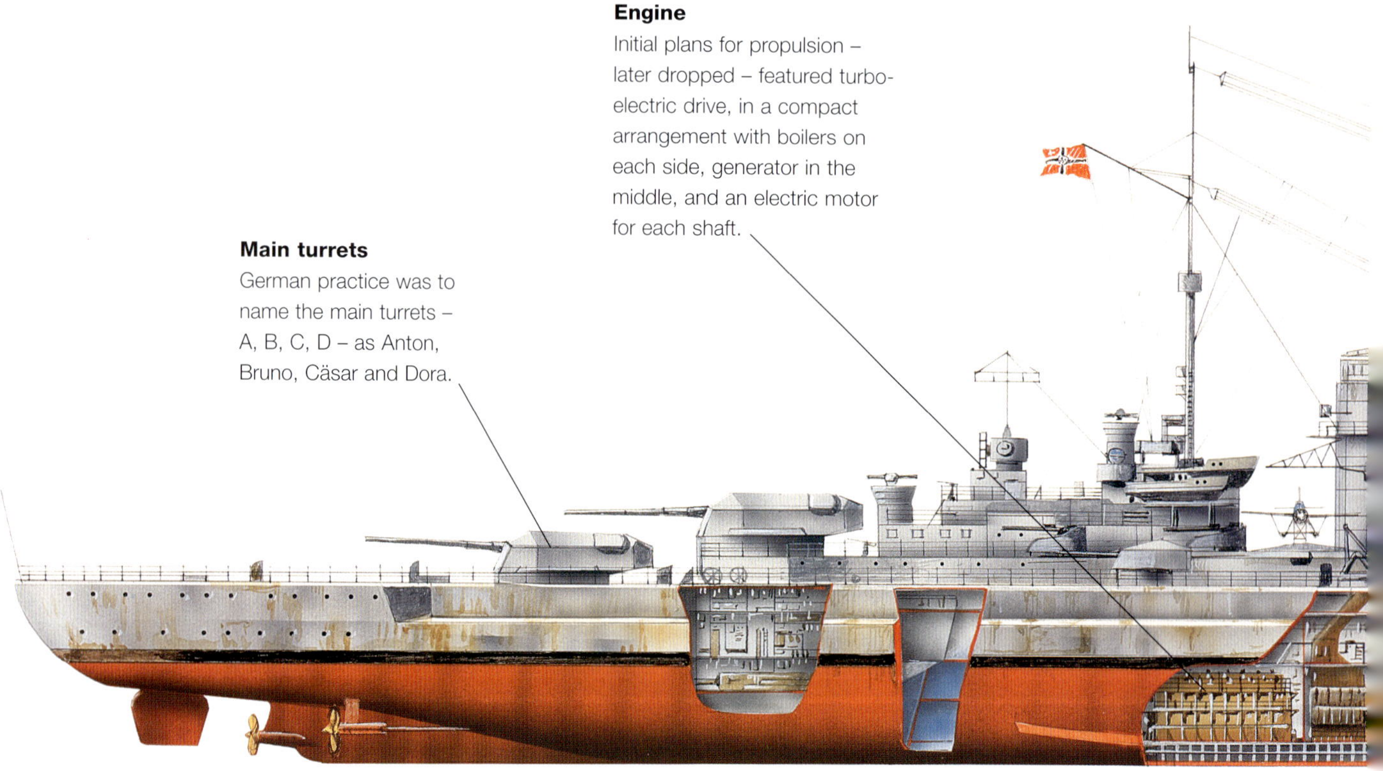

shortening shell flight time). The shells weighed 798kg (1759lb) and were of three kinds: armour-piercing HE for use against battleships and two types of HE shell, one with a head fuse and one with a base fuse. Identical guns were fitted on *Tirpitz* – 12 140mm (6in) guns fitted in twin turrets, three on each side. AA defences consisted of 16 105mm (4.1in) twin-mount guns, 16 37mm (1.5in) twin-mount and 12 20mm (0.79in) single-barrel guns.

Gunnery control

Gunnery control was located in three positions, on the forward conning tower, above the foretop platform and on the rear conning tower, with communication links to two computation rooms, within the armoured zone fore and aft. Unlike *Tirpitz*, *Bismarck* carried no torpedoes but both ships had four Arado Ar196 floatplanes. The hull was constructed from ST52 steel and more than 90 per cent was of electric welding construction. Armour was of three kinds, Krupp KCn/A face-hardened for the belt, turrets and control towers; Wh – *Wotan hart* ('hard') – for the armoured decks, and Ww – *Wotan weich* ('soft') – for the longitudinal torpedo bulkheads.

Armour displacement

Altogether the armour weight was 15,654 tonnes (17,256 tons), 32.2 per cent of total displacement. The hull was divided into 22 watertight compartments, 17 within the armoured zone which accounted for 70 per cent of the

The cutaway plan gives some idea of the extent and subdivision of internal compartments within the hull. This arrangement was one of prime reason's for the ship's remarkable staying-power in combat.

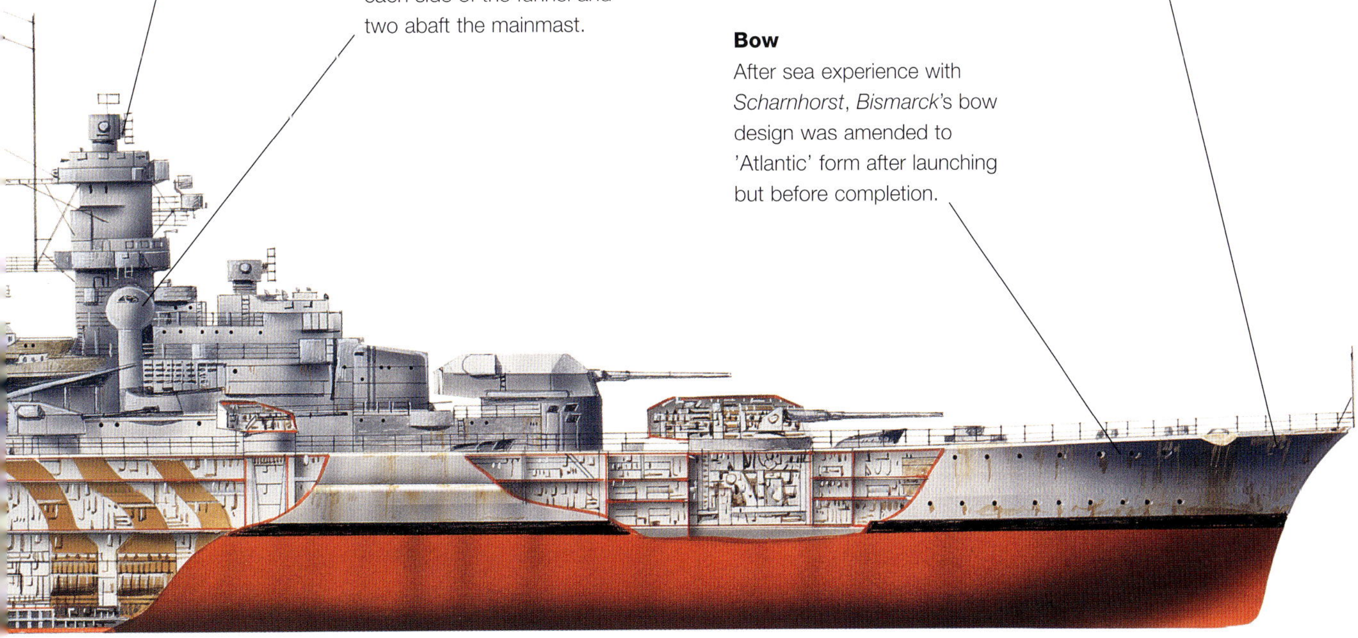

Radar equipment
Bismarck carried FuMO radar equipment on the forward and aft rangefinders and the foretop, though blast concussion disabled it. It also had hydrophone detectors.

Searchlights
Seven 1.5m (59in) diameter searchlights were carried, one on the foremast, four each side of the funnel and two abaft the mainmast.

Bow
After sea experience with *Scharnhorst*, *Bismarck*'s bow design was amended to 'Atlantic' form after launching but before completion.

Bow anchors
Three bow anchors were carried, two on the port side and one stern anchor (also port side). Each anchor weighed 9500kg (20,944lb).

Specification

Dimensions	Length 251m (793ft), Beam 36m (118ft), Draught 9.3m (31ft), Displacement 37,194 tonnes (41,000 tons); 44,905 tonnes (49,500 tons) full load
Propulsion	12 Wagner HP boilers, 3 Brown-Boveri geared turbines developing 111,982kW (150,170hp)
Armament	8 380mm (15in) guns, 12 150mm (5.9in) guns, 16 105mm (4.1in) guns, 16 37mm (1.5in) guns, 12 20mm (0.79in) AA guns
Armour	Belt 320–80mm (12.6-3.1in), Bulkheads 220–45mm (8.6–1.7in), Deck 120-80mm (4.3–1.7in), Barbettes 340–220mm (13.4–8.6in), Turrets 360–180mm (14.2-7in)
Range	16,430km (8870nm) at 19 knots
Speed	29 knots
Complement	2092

Bismarck photographed from _Prinz Eugen_, en route to the Atlantic, 19–20 May 1941. The camouflage paint represents a false bow and stern, to minimise the ship's great size.

ship's waterline length. From foretop to keel were 17 levels or decks. In total 12 high-pressure boilers in six watertight compartments supplied steam to three Blohm & Voss turbine sets, each in separate rooms and driving the three screws. Eight 500kW diesel generators, five 690kW and one 460kW turbo-generators were installed in four plants. In terms of internal layout, levels of protection and capability in action, this was the most effective battleship yet built.

Under the name of Operation Rheinübung, *Bismarck* and *Prinz Eugen* left Gdynia on 18 May 1941 to raid merchant shipping routes in the Atlantic, with Admiral Günther Lütjens in overall command. It was a carefully planned mission, with supply ships and scouting U-boats in position. News of the departure reached the British from Sweden and from Norwegian resistance observers, and cruisers were deployed to keep watch while an interception force was mobilized.

On 24 May the battlecruiser HMS *Hood* and battleship *Prince of Wales* engaged the German ships, and *Hood* was sunk by *Bismarck*'s fifth salvo. Hits by three 356mm (14in) shells from *Prince of Wales* had done comparatively little damage to *Bismarck*. A huge three-day search and pursuit followed, with some 48 ships involved, including five battleships, two aircraft carriers, nine cruisers and 18

Adolf Hitler inspects the almost-completed *Bismarck* at the Gotenhafen (Gdynia) naval base on 5 May 1940.

destroyers. *Prinz Eugen* reached Brest safely but *Bismarck*, after almost succeeding in dodging the pursuit, was hit by Swordfish torpedo bombers from HMS *Ark Royal* at 20:47 on 26 May, jamming the port rudder at 120 degrees to port.

Steaming through the night at only 18.5km/h (10 knots), *Bismarck* kept up a vigorous defence against by destroyers and from the air. On 27 May at 08:47 the battleships HMS *Rodney* and *King George V* opened fire at around 20,000m (21,900yd), gradually closing to 7860m (8600yd). By around 10:00 all *Bismarck*'s guns were disabled and the order was given to scuttle. Hit by more British torpedoes, it capsized and sank at 10:39 with 1977 of the 2221 people on board.

Bismarck resilient

Despite the prolonged and ferocious pounding by 408mm (16in) and 356mm (14in) guns and numerous torpedo strikes, neither the armoured belt nor the armoured deck of *Bismarck* had been penetrated, apart from a single 356mm (14in) shell that penetrated the side armour. The one fatal strike was that of the aerial torpedo which jammed the port rudder, preventing the ship's run for port. With the ammunition exhausted and its guns out of action, the crew were preparing for the ship's self-destruction when three torpedoes from the cruiser HMS *Dorsetshire* struck the armoured hull. At first the British believed that these had finally sunk *Bismarck*, but all the battleship's internal equipment continued to operate, including the engines. Had the war begun in the year Hitler had first envisaged, allowing another five years of German naval preparation resulting in a further six battleships of *Bismarck* standard, the war at sea would have been very different.

Vittorio Veneto (1940)

Italy s last battleship class was to be formed of four ships, but only three were completed, of which two saw active service. Despite new and ingenious armour protection they proved vulnerable to airborne attack.

Classified officially as *corrazzate* (literally 'armour-clads'), the class design goes back to 1928. *Vittorio Veneto* was laid down at Cantieri Riuniti dell'Adriatico in Trieste on 28 October 1934, launched on 25 July 1937 and completed on 28 April 1940. Originally intended as a 31,751 tonne (35,000 ton) ship under the Washington Treaty, the design was enlarged after France refused to ratify the Agreement in 1935 and recommenced battleship construction.

Construction problems

The lengthy construction time reflects difficulties in the provision of materials and equipment, and much of the build quality was poor. However the general appearance was both warlike and elegant. For a time 406mm (16in) guns were proposed but were rejected in favour of 381mm (15in) as a design for these was already available, having been prepared for the cancelled *Francesco*

Rear turret
The high placing of the rear main turret was in order to keep blast effects clear of the aircraft deck.

Aircraft
Ro 43 reconnaissance aircraft were carried. Littorio in 1942 carried floatless Reggiane 2000 fighters that could be launched from the ship but had to find a landing ground on shore.

One sister ship, *Littorio*, renamed *Italia* after Italy's surrender in 1943, was scrapped in 1952–54. The other, *Roma*, was sunk by German 'Fritz X' radio-controlled glide bombs on 9 September 1943.

Radar aerial

From 1943 Vittorio Veneto carried a radar aerial on the turret mast.

Stems

Vittorio Veneto and Littorio were launched with slightly raking stems, giving a waterline length of 236m (774ft 3in). After trials, they were given straight stems down to the waterline, resulting in an increase of 1.8m (4ft 6in).

Various styles of camouflage paint were applied during *Vittorio Veneto*'s wartime action, depending on season and mission. Here it is probably at Taranto in 1940.

Specification

Dimensions	Length 237.8m (778ft 9in), Beam 32.9m (107ft 9in), Draught 9.6m (31ft 5in), Displacement 37,536 tonnes (41,377 tons); 41,505 tonnes (45,752 tons) full load
Propulsion	8 Yarrow boilers, 4 Belluzzo geared turbines developing 100,383kW (134,616hp), 4 screws
Armament	9 381mm (15in) guns, 12 152mm (6in) guns, 4 120mm (4.7in) guns, 12 90mm (3.5in) AA guns, 20 37mm (1.4in) and 32 20mm (0.79in) AA guns
Armour	Belt 350–60mm (13.8–2.4in), Bulkheads 100–70mm (3.9–2.75in), Deck 205–35mm (8.1–1.4in), Barbettes 350mm (13.8in), Turrets 350–100mm (13.8–3.9in)
Range	7400km (4000nm) at 16 knots
Speed	31.4 knots
Complement	1861

Carraciolo class battleships. Nine main guns were fitted in triple turrets. With a maximum elevation of 35 degrees they had a range of 42,260m (46,210yd) and fired both AP (armour-piercing) shells of 885kg (1950lb) and HE shells of 774kg (1710lb). They fired a round every 45 seconds.

Gun calibre was 50, and muzzle velocity was very high – 870m (2854ft) per second – and this led to overstated claims about the penetrative impact. But its disadvantages included both wear on the barrel lining and the tendency of shells to disperse instead of following the intended trajectory. Secondary armament was 12 152mm (6in) guns in triple turrets, plus four 120mm (4.7in) guns. Substantial AA defences were incorporated from the start, with 12 90mm (3.5in), 20 37mm (1.5in) and 16 20mm (0.79in) guns providing both long-range and close-in defence.

Total armour weight was 12,093 tonnes (13,331 tons) and protection included massive side armour with further internal bulkheads, intended to counter the impact of plunging shells. Against torpedoes it had the Pugliese system: a lateral space between the torpedo bulkhead and the inner hull bulkhead, holding a long empty cylinder, maximum diameter 3.8m (12ft 6in), suspended in oil, and intended to absorb explosive energy. It was the first Italian battleship to have radar installed: a 'Gufo' EC4 device fitted in 1942.

The role of an Italian capital ship in the late 1930s had much to do with rivalry with France, the only other Mediterranean power building new battleships. (Franco's Spain, after 1939, flirted with the idea of acquiring one or two on the *Vittorio Veneto* model, but did not pursue it.) In reality, the Mediterranean was controlled by the British Navy, whose task it was to keep open the short route to Suez and on to India. Ironically, when Italy for a time broke this British hegemony, it was not by use of battleships but by the smallest possible craft, midget submarines (see *Queen Elizabeth*).

British confrontations

Assigned to the 9th Division of the 1st Squadron from 31 August to the end of 1940, *Vittorio Veneto* was frequently in action against British convoys to Malta, but without success. It survived the raid of 12 November on Taranto and was transferred to Naples, from where it continued to make sorties. It was the Italian flagship in the battle of Cape Matapan, 27 March 1941, when it was hit by a torpedo launched by a bomber from HMS *Formidable*. Taking on around 3538 tonnes (3900 tons) of water, it struggled back to Taranto.

Restored to duty at the beginning of August 1941, the ship hunted further Malta convoys and also escorted Italian convoys to North Africa. On 14 December it was hit by a torpedo from the British submarine *Urge* and was out of action until June 1942. Its last active service, again against Malta convoys, was in that month. In all it took part in 56 war missions, of which 11 were conducted against enemy ships.

On 8 September 1943 Italy obtained an armistice with the Allies and its fleet was surrendered. *Vittorio Veneto* was moved first to Malta, then to the Great Bitter Lake in the Suez Canal for internment. It was returned to Italy in February 1946 and allocated to Great Britain as reparation. Decommissioned on 1 February 1948, it was scrapped at La Spezia between then and 1950.

Torpedo attacks

By 1940 the aerial torpedo was a lethal weapon against any battleship that was not very heavily protected both against air attack and from an underwater hit. In November 1940 three Italian battleships moored in Taranto harbour were attacked by British torpedo-carrying aircraft (see picture right). *Littorio*, newly completed, was struck by two torpedoes and began to sink immediately, but the captain succeeded in running the ship on to the beach shortly before a third torpedo hit (it was refloated and repaired). *Caio Duilio*, an older ship (1915) but which had been extensively modernised, succumbed to a single torpedo hit and was towed into shallow water where it sank. *Conte di Cavour*, another battleship that had been refitted for modern warfare, also received only a single hit, which was enough to sink it immediately. The torpedoes used were relatively small calibre, 432mm (17in), with an explosive force only 70 per cent of that of a ship-fired 533mm (21in) torpedo.

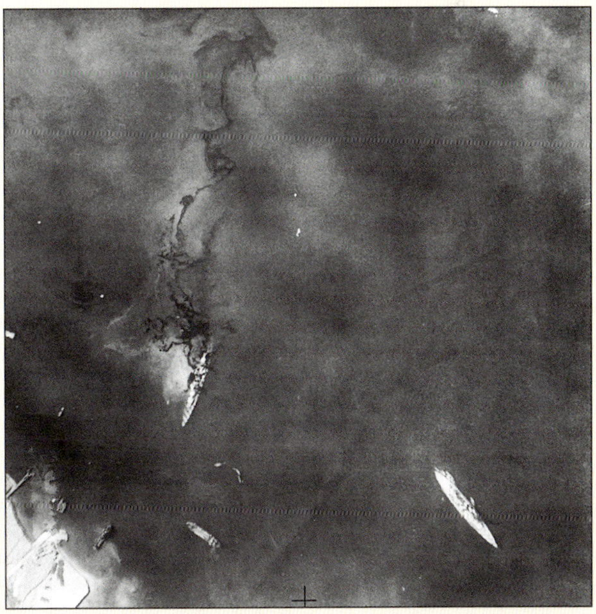

Richelieu (1940)

The newly-commissioned *Richelieu* was attacked by Allied ships after the fall of France but later was taken over by Free French forces, and played a part in the struggle against the Axis powers both in European and Far Eastern waters.

The *Dunkerque* class of 24,040 tonne (26,500 ton) battleships (1931) was the model for three new and larger ships of the *Richelieu* class. Only one, *Richelieu*, was fully completed. Laid down at the Arsenal de Brest on 22 October 1935, it was launched on 17 January 1939 and completed on 15 June 1940, a week before the fall of France. The cost was 1,227 million francs.

Like the British battleships *Rodney* and *Nelson*, *Richelieu* carried all its main armament forward. These were eight 380mm (15in) guns mounted in two quadruple turrets, with 'B' turret superfiring. The guns were cradled in pairs, and had a range of 45,000m (49,210yd) at an elevation of 35 degrees. The HE shells weighed 880kg (1940lb) and rate of fire was two rounds a minute.

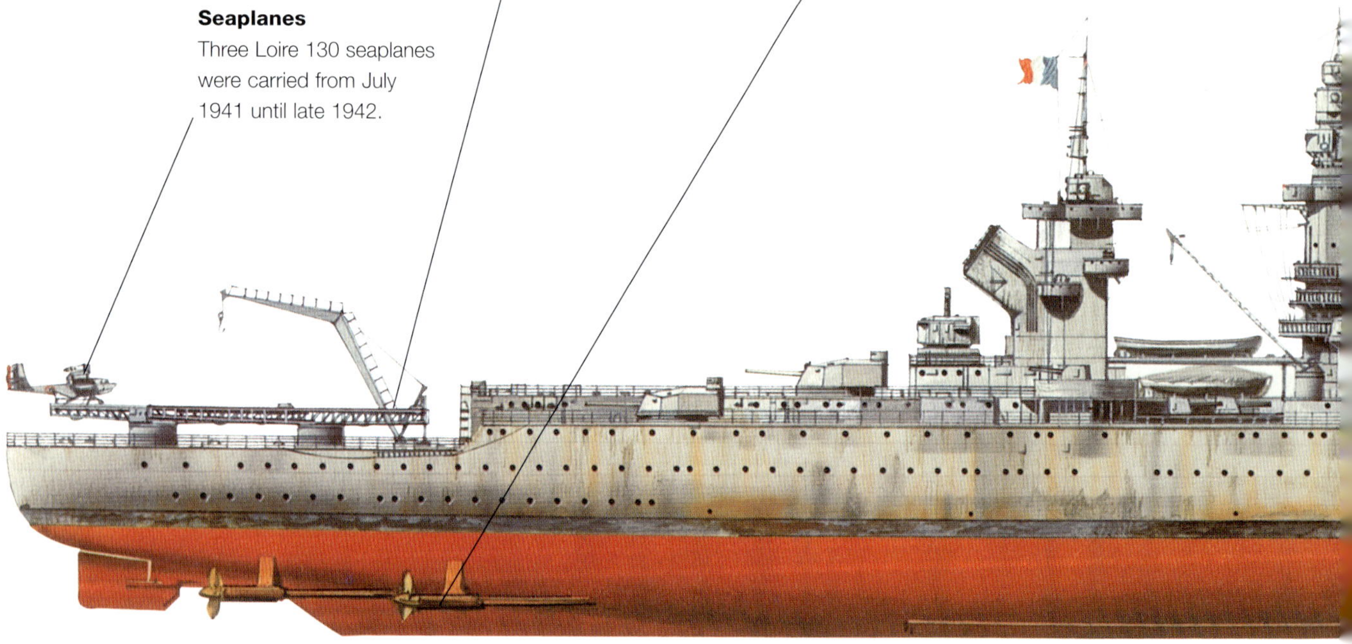

Aircraft deck
The aircraft deck was used to house an AA battery after the 1943 repairs in New York.

Propeller shafts
Both starboard shafts were severely damaged in the air attack of 8 July 1940.

Seaplanes
Three Loire 130 seaplanes were carried from July 1941 until late 1942.

Secondary armament was nine 152mm (6in) guns in triple turrets (15 had originally been intended), 12 100mm (3.9in) AA guns and 16 37mm (1.4in) AA guns, all in twin mountings. From the American refit of 1943 the 37mm guns were replaced by 56 40mm (1.6in) Bofors guns in quadruple mounts. With only between five and seven battleships in commission through the 1930s, the French Navy saw their role as swift-striking raiders, despite the designation of *bâtiment de ligne*, and *Richelieu* was engined for 55.5km/h (30 knots) and attained a maximum of 59.2km/h (32 knots).

During construction numerous changes were made to the original design which had the funnel placed between the

Quadruple turrets
Fire from the quadruple turrets was improved from 1948, when retarders were fitted to the outer guns, giving them a delay of 60 milliseconds – enough to avoid mutual dispersional effects of simultaneously-fired shells.

After Far East service, *Richelieu* returned to liberated France on 11 February 1946. Transferred to reserve status in 1958, it was not finally stricken from the Navy list until January 1968.

Boilers
Richelieu's new type pressure-fired boilers supplied the highest power rating of any battleship until the US *Iowa* class in 1943.

Specification

Dimensions	Length 247.9m (813ft), Beam 33m (108ft), Draught 9.7m (32ft), Displacement 39,275 tonnes (43,293 tons); 43,135 tonnes (47,548 tons) full load
Propulsion	6 Indret pressure-fired boilers, 4 Parsons geared turbines developing 111,855kW (150,000hp), 4 screws
Armament	8 380mm (15in) guns in quadruple mounts, 9 152mm (6in) guns, 12 100mm (3.9in) AA guns
Armour	Belt 343–250mm (13.5–9.8in), Bulkheads 383–251mm (15.1–9.9in), Deck upper 170–130mm (6.7–5.1in), Deck lower 100–40mm (3.9–1.6in), Barbettes 405mm (16in), Turrets 430–195mm (16.9–7.7in)
Range	18,240km (9850nm) at 16 knots
Speed	30 knots
Complement	1550

main and aft towers. The funnel and aft tower were merged into a single structure, with the funnel angled sharply towards the stern. Despite the ever-increasing size and displacement of capital ships, designers were always looking for possible ways to save weight, if only to reuse it elsewhere, and this was one solution.

Richelieu was well-armoured, with 14,878 tonnes (16,400 tons), amounting to 37 per cent of displacement, concentrated in the area between the fore turret and the aft 152mm (6in) turret. The long forecastle was almost unprotected, though a 40mm (1.6in) armoured deck was located below the design waterline.

The after deck was given over to a hangar and two catapult sets, for three aircraft, with a hoisting crane. In 1941 a French surface/air radar system, *Directeur electro-magnétique* (DEM), was fitted, with two emission antennae on the foremast and two reception antennae on the after tower.

British attack

Richelieu was moved from Brest to the Dakar base on 18 June 1940, before engine trials were completed. On 8 July, as a potential enemy, it was attacked by Fairey

Ammunition problems

From 1943 *Richelieu* as France's largest battleship was a powerful symbol of Free France while the country lay under German occupation. But as a fighting ship it had serious problems. It had left France short of ammunition and could not get more shells from French factories. The Americans had no ships with 380mm (15in) guns and so had no suitable shells. There were also problems with the propellant charges stored at the Dakar base. The damaged guns were replaced by three from the sister ship *Jean Bart*, which though only 77 per cent completed, had steamed in June 1940 from St Nazaire to Casablanca, but had been immobilised there later by attacks from US forces. It was May 1944 before *Richelieu* was supplied with suitable ammunition from British factories and could take an effective part in shore bombardment. *Jean Bart*'s three remaining guns, plus one newly-made, were installed in *Richelieu*'s 'A' turret in 1951, while the original guns from that turret replaced those in 'B' turret.

Richelieu in early 1946, probably at Singapore and about to sail home, still with wartime hull paintwork, intended to disguise the true length of the ship.

Swordfish aircraft from the British carrier *Hermes* and sustained further damage in shell hits from the *Revenge* class battleship HMS *Resolution* while fending off an Allied attack on 23–25 September. Premature explosion of its own shells wrecked three of the four guns in the superfiring turret. Makeshift repairs continued at Dakar, until following the Allied landings in North Africa (November 1942), French forces joined the Allies, and in January 1943 *Richelieu* was sailed to New York, with catapults and aircraft removed, for full repairs.

Deck plan, showing the original pre-1943 configuration, with aircraft catapults at the stern. From September 1943 these were replaced by AA batteries of 4-barrel 40mm (1.6in) Bofors guns, and 20mm (0.79in) Oerlikon AA guns.

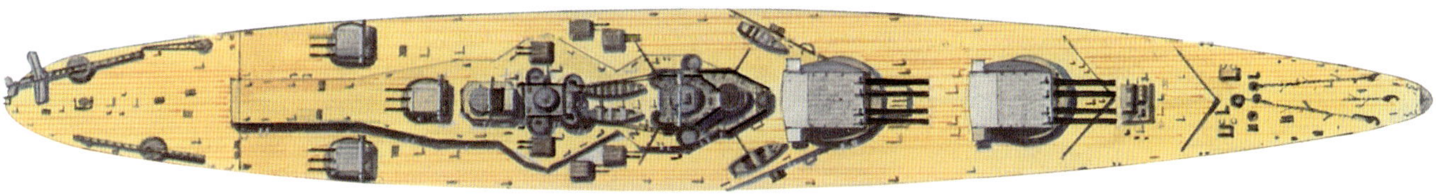

Yamato (1941)

The class were to be 'super-battleships', bigger, more heavily armed and better-protected than anything else afloat. Intended to enforce Japan's mastery of the Pacific, they made a minimal contribution to the country's war effort.

Work on the *Yamato* class began, in secret, in 1934 and three years of intensive planning and modelling followed. *Yamato* was laid down in the Kure Kaigun Kosho building basin on 4 November 1937, launched on 8 August 1940 and completed on 16 December 1941, as the first of a planned five that were intended to be the largest and most powerful battleships in the world.

The nine 460mm (18.1in) 45-calibre guns were the heaviest ever used afloat. The triple turrets each weighed 2516 tonnes (2774 tons). Barrel length was 21.13m (69ft

4in), they weighed 162 tonnes (178.6 tons) and had a range of 44km (27.3 miles) at an elevation of 45 degrees. The HE shells weighed 1460kg (3219lb). The secondary armament was 12 155mm (6.1in) guns mounted in four triple turrets and 12 127mm (5in) guns in twin mounts. Building was done on the raft body principle, with the vital areas contained within side armour of 410mm (16in) thickness, tapering towards the bottom to 75mm (2.9in), topped by a 200mm (7.8in) armoured deck and terminated by transverse bulkheads. Only the barbettes, flue gas uptakes and trunks for command systems, all heavily armoured, protruded through the 'raft'.

The tower carried Type 21 and 22 radar, the main battery range-finder, and Type 98 low-angle fire control director. Uppermost bridge deck is the combat bridge, with compass bridge below. The conning-tower top with its periscopes is just above the 155mm (6.1in) turret.

Armoured deck
The armoured deck was designed so that it could only be pierced by bombs of 1,000kg (2,200lb) or more, dropped from a height exceeding 2,400m (7,800ft).

Hull
The hull shape was carefully worked out to bear the huge stresses. The stem design was estimated to reduce water resistance by 8.2 per cent, and the overall design gave a power saving of 11,797kW (15,820hp).

Yamato at speed, photographed from an American warplane while under bomb attack in the Leyte Gulf sea battle, 25 October 1944.

Underwater defences

Torpedo protection consisted of a bulkhead and a torpedo bulge with a maximum width of 3m (10ft), and to guard against explosions from below the side armour was continued as a floor 80–50mm (3.15–2in) thick beneath the magazines, with a space of around 4m (13ft 6in) to absorb explosive energy. Altogether there were 1065 watertight compartments below the armoured deck, and 82 above.

Instead of the 'pagoda' style superstructure of previous Japanese battleships there was a tall octagonal tower-mast, reaching 28m (92ft) above the waterline, with relatively few external features, though signalling wings were built out at bridge level. Control and chartrooms were arranged round a central armoured cylinder. A 15m (49ft) rangefinder surmounted the tower, with gunnery control centres above and below. The upper bridge extended forwards some 5m (16ft 4in), flanked by triple searchlights on each side.

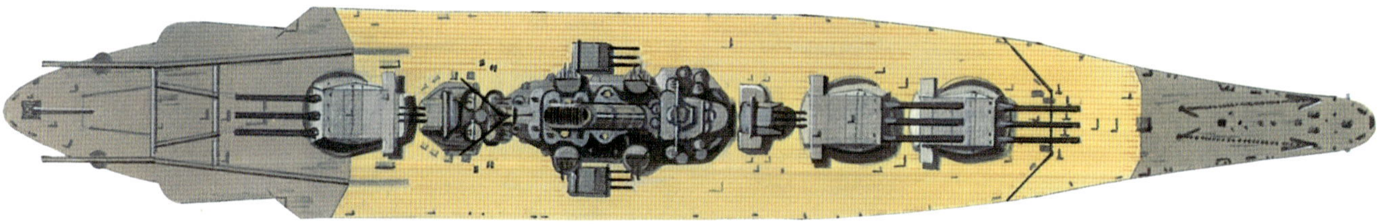

The deck plan reveals the distinctive hull shape, reaching maximum beam towards the stern. The 'wings' carried Type 96 25mm AA guns in triple mounts.

Kampon line

The hull attained its maximum beam aft of the mid-point, part of the design scheme sometimes referred to as the 'Kampon line' and intended to minimise the stresses caused by the ship's great length and the massive weight of the turrets. Despite its huge dimensions, *Yamato* was intended to be a fast ship, and at one stage diesel propulsion was proposed for the two outer shafts, with turbines for the inner ones. In the end an all-turbine drive was chosen, as in the original plan.

After trials and training exercises, *Yamato* served as flagship of the Japanese Combined Fleet, and as such was command ship in the Battle of Midway at the end of May 1942, in which four Japanese fleet carriers were sunk. *Yamato* was not involved in the fighting. By August 1942 its sister ship *Musashi* had been completed and replaced *Yamato* as flagship on 11 February 1943. *Yamato* moved between the mid-ocean base of Truk (Chuuk) and the home base of Kure, where it underwent a refit in July–August.

On 25 December, *Yamato* was torpedoed by the US submarine *Skate* while it was carrying troops and equipment from Yokosuka to Truk. The attack caused the rear magazine to flood when some 2721 tonnes (3000 tons) of water surged in. Repairs and refitting at Kure lasted until 18 March 1944. Between 22 and 25 October 1944 it was part of the Japanese Centre Force, engaged in the widespread Battle of Leyte Gulf. *Yamato*'s guns sank the US escort carrier *Gambier Bay* and it participated in the sinking of three US destroyers. It received several bomb and shell hits without sustaining serious damage, but *Musashi* was sunk by US aircraft. *Yamato* returned to Kure.

Specification (1941)

Dimensions	Length 263m (862ft 9in), Beam 36.9m (121ft 1in), Draught 10.39m (34ft 1in), Displacement 61,698 tonnes (68,010 tons); 65,008 tonnes (71,659 tons) full load
Propulsion	12 Kampon HP boilers, 4 Kampon turbines developing 111,855kW (150,000hp), 4 screws
Armament	9 460mm (18.1in) guns, 12 155mm (6.1in) guns, 12 127mm (5in) guns, 24 25mm (0.98in), 4 13.2mm (0.52in) AA guns
Armour	Belt 410mm (16in), Deck 230–200mm (9.1–7.9in), Barbettes 546–50mm (21.5–2in), Turrets 650–193mm (19.7–11.8in), Torpedo Bulkhead 300–75mm (11.8–2.9in)
Range	13,330km (7200nm) at 16 knots
Speed	27 knots
Complement	2500

Successful class?

Japan's most able naval designers, engineer Hiraga and Captain Fujimoto, made major contributions to the design of the class, which was generally recognised as a highly successful and effective one despite going beyond all previous bounds of size. However, none of the *Yamato* class achieved results comparable to their size, expense and power. *Musashi* was sunk by aerial bombs and torpedoes. The third ship, *Shinano*, was converted while building to a carrier. Newly completed, it was sunk by torpedoes from the US submarine *Archerfish* on 29 November 1944. Construction of the fourth ship, never named (No.111), was suspended in November 1941 when it was about 30 per cent completed, and finally abandoned in September 1942. A fifth had been envisaged but no construction order was placed.

'Suicide mission'

On 6 April 1945 it was sent to help repel the American landings on Okinawa in an operation code-named Ten-go, generally considered a large-scale suicide mission. *Yamato* was to be beached on the island to act as a fixed artillery fortress. With nine escorting craft but no air protection, it was attacked on the 7th, southwest of the Kyushu Islands, by around 400 American bombers and torpedo bombers in three waves. The attacks began at 12:37 and, hit by six bombs and 11 or more torpedoes, *Yamato* was progressively disabled and partially flooded, with little power and no steering. At 14:23 the ship capsized, one of the two fore magazines exploding at the same time. Around 2055 of the crew were killed or drowned.

Prince of Wales (1941)

The second *King George* V class battleship, *Prince of Wales,* had a brief career: sunk off Malaya by aerial attack less than a year after commissioning, it gained the unfortunate distinction of being the first battleship sunk by aircraft on the open sea.

On 31 December the Washington Treaty ended and was not renewed. But plans had gone ahead in anticipation. In the USA and Japan, 408mm (16in) guns were being assembled for battleships. Britain had no blueprints for guns of this calibre, but did have a 356mm (14in) quick-firing gun of excellent quality, produced in the early 1930s, and the first of its type to be built in the form of a jacketed and hooped barrel.

The *King George V* class, emerging from the 1936 building programme, was something of a compromise design. Based on a 31,751 tonne (35,000 ton) hull, it was to carry 12 356mm (14in) guns in three quadruple turrets. During design, to save weight for use elsewhere, the superfiring 'B' turret was reduced to a twin mounting. All five were laid down in 1937, *Prince of Wales* at Cammell Laird's yard in Birkenhead in January.

Aircraft
Four Supermarine Walrus seaplanes were originally carried by the *King George V* class. The hangars were used as cinema and dressing stations from 1943.

Turrets
Weight of the quadruple turrets was 1383 tonnes (1525 tons).

Launched on 3 May 1939, *Prince of Wales* was completed in March 1941. The design was a balanced and handsome one, with the fore-funnel marginally taller than the after one; and the first British battleship to have quadruple main turrets. Apart from *Rodney* and *Nelson*, these were the first British battleships built since World War I, and they were seen as the new – and vitally necessary – backbone of the RN in a new conflict with Germany.

Preparing for war

The advent of war prompted numerous alterations during completion, including extra ammunition, fuel capacity and radar antennae, increasing design displacement to

The arrangement for accommodating aircraft, with a hangar each side of the forward funnel, and rails to a catapult set athwartships, begun with HMS *Warspite* in the mid-1930s. It was admired by the German Navy, which adapted it for the *Bismarck* class.

Prince of Wales

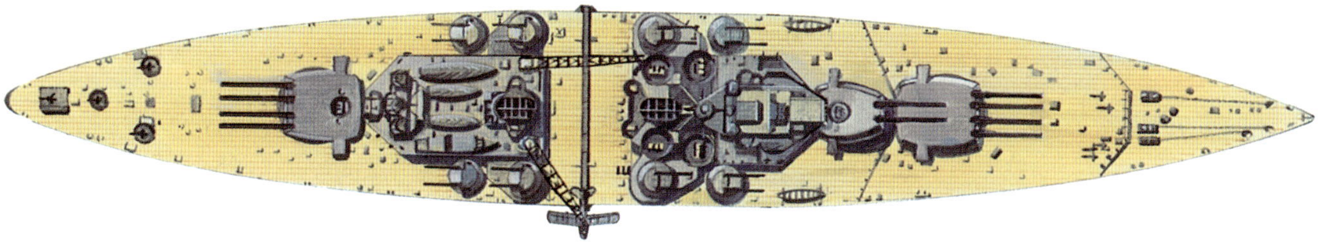

Deck plan: of the five ships in the *King George V* class, only two, the lead vessel (up to 1943) and *Prince of Wales* carried aircraft, though the superstructure gap was intended for their deployment. It may also have helped in overall balancing of a hull 227m (745ft) long.

34,473 tonnes (38,000 tons) and deepening the draught from designed as 8.4m (27ft 8in) to 9.75m (32ft) with a corresponding reduction in freeboard.

Secondary armament was 16 133mm (5.25in) guns, fully automatic and with a range of 20,570m (22,500yd), capable of long-range AA defence as well as knocking out destroyers. Original plans allowed for 32 40mm (1.6in) guns in eight-barrel mountings, and 16 four-barrelled heavy machine guns, but the machine guns were not installed on *Prince of Wales*. It was, however, fitted with eight UP smooth-bore multi-barrel AA batteries, ingenious but not very effective weapons intended to dangle small aerial mines from parachute wires.

Eight boilers in four compartments provided steam to four independent sets of geared Parsons turbines. The design speed could be forced up to 54km/h (29.2 knots) with an output of 93,212kW (125,000hp). The class carried approximately 10,886 tonnes (12,000 tons) of armour, with a high-level armoured deck resting on top of the side armour, with a maximum thickness of 152mm (6in) above the magazines. Protection was on the 'all or nothing' principle, and armour amounted to 40 per cent of design displacement. Careful design of the engine space made it possible to limit the length of the central armoured area to 136m (446ft) on a waterline length of 227m (744ft 9in). Torpedo bulges were not fitted, but the torpedo bulkhead was 51mm (2in) and effective torpedo protection space was 4.1m (13ft 4in).

Combat mission

Prince of Wales was still engaged on trials when ordered out on its first combat mission with HMS *Hood* against *Bismarck* and *Prinz Eugen* on 24 May 1941. An early hit knocked out the gunnery control system and it suffered considerable damage, withdrawing from the battle after the sinking of *Hood*. In August 1941 it carried Winston Churchill to meet President Franklin D. Roosevelt at Argentia Bay, Newfoundland. In September it was with Force 'H' in the Mediterranean, and after a brief return to Britain was deployed to the Far East, arriving in Singapore on 27 November 1941.

Specification

Dimensions	Length 227.1m (745ft 1in), Beam 31.4m (103ft 2in), Draught 10.5m (34ft 4in), Displacement 40,415 tonnes (44,550 tons) full load
Propulsion	8 Admiralty 3-drum boilers, 4 Parsons geared turbines developing 82,000kW (110,000hp)
Armament	10 360mm (14in) guns, 16 133mm (5.25in) guns, 32 40mm (1.6in) AA guns, 80 UP projectors
Armour	Belt 381–114mm (15–4.5in), Bulkheads 381mm (15in), Deck 152–127mm (6–5in), Barbettes 330–280mm (13–11in), Turrets 330–152mm (13–6in)
Range	28,900km (15,600nm) at 10 knots
Speed	28.3 knots
Complement	1612

On 8 December *Prince of Wales* fired its AA guns against Japanese aircraft over Singapore. On 10 December 1941, it was off the east Malayan coast as flagship of Force 'Z' with the battlecruiser *Repulse* and four destroyers when they were attacked by successive waves of (land-based) Japanese naval aircraft armed with bombs and torpedoes. The ships opened intensive AA fire, increased speed and began zig-zagging, but control of *Prince of Wales*'s port propellers was lost and its manoeuvrability was restricted.

After numerous bomb and torpedo hits, both ships were sunk. The escorting destroyers rescued 1285 men from *Prince of Wales*. The other four ships of the class, *King George V*, *Anson*, *Howe* and *Duke of York*, survived the war and were for a time considered as possible guided-missile carriers, but all were scrapped in 1957.

This undated photograph may show the ship on its return from the engagement with *Bismarck.* The British style of bridge design, begun with *Rodney*, is quite different from that of the German battleships.

Air defences

The loss of two capital ships – one still new – to Japanese air attack raised many questions in Britain about their anti-aircraft defences and also about the viability of battleships in this war. The main criticism of the class had been the comparatively low calibre of its main guns, but this was not an issue in the loss of *Prince of Wales*. The tactics of Admiral Phillips, in command of Force 'Z', were queried, especially his failure to request available air support.

Although *Prince of Wales* carried far less AA defence than major ships would do later in the war, it was probably regarded at the time as well-equipped to deal with air attack through its High Angle Control System, linked to long-range air-search radar.

From July 1941 it was carrying radar of Types 279 (air warning), 284 (surface), 282 and 285 (AA control), though on 10 December 1941 the surface scanning radar was inoperable.

Tirpitz (1941)

Sister ship to *Bismarck, Tirpitz* had similar qualities of power and resistance. Although its active service was very limited, the fact of its presence tied up substantial British resources until its ultimate destruction.

Tirpitz was laid down on 26 October 1936 at the Wilhelmshaven Naval Dockyard, launched on 1 April 1939 and completed on 25 February 1941. Its cost was 181.6 million Reichsmarks, 15.2 million less than *Bismarck*. It carried the same main armament, eight 380mm (15in) guns in four twin turrets, capable of firing three rounds per minute, 12 150mm (5.9in) guns in six twin turrets, 16 105mm (4.1in) and 16 37mm (1.4in) twin-mount AA guns. *Tirpitz* originally had 12 20mm (0.79in) single-barrel AA guns; this was reduced to 10, but supplemented by 40 20mm four-barrelled guns. From 1942 it also carried eight 533mm (21in) torpedo tubes in sets of four on the upper deck. Up to six floatplanes could be carried, and a double catapult was fitted.

Telecommunications refit

The superstructures of both ships were very similar, with a foremast stepped on an upper level of the tower and a single large funnel. Numerous modifications were made in the course of 1941–44 to accommodate new radar and

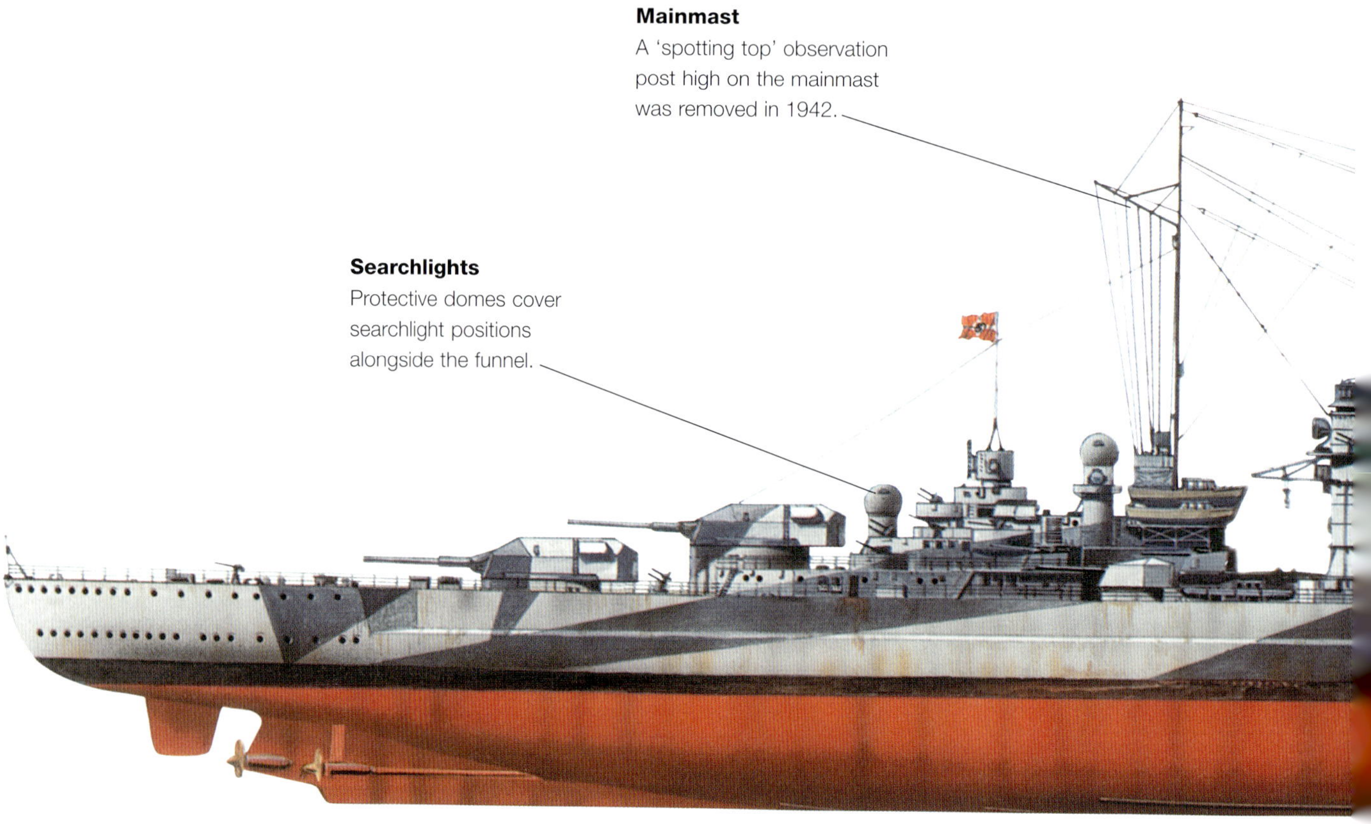

telecommunications equipment as well as extending the AA defences. Turbo-electric drive had been planned originally, since it had proved successful in fast ocean liners, but through doubts about its efficacy on a battleship the designers fell back on extra-high pressure boilers and steam turbines as in the *Scharnhorst* class.

Radar, at first able to indicate range but not height, was installed with a large aerial placed on the forward face of the rotating range-finder dome on the ship's fore-top. Only *Tirpitz* and *Bismarck* carried a radar aerial fixed on the rotating dome above the command post. From 1941–42 a similar aerial was also placed on the after dome. In

Tirpitz is depicted below in 'Paint Scheme K' which it carried between March and July 1944. The camouflage style was frequently changed to confuse aerial observers.

Funnel
At first black, the funnel cowling was painted silver- or light-grey from 1942. The catwalk from funnel platform to tower was added in 1942.

Cranes
The cranes were fitted one deck higher and slightly further forward than in *Bismarck*.

Superstructure
The open fore part of the superstructure, hardly higher than 'B' turret, was rimmed out to deflect blast from gunfire.

Stem
The revised 'Atlantic' stem was fitted to *Tirpitz* before launching.

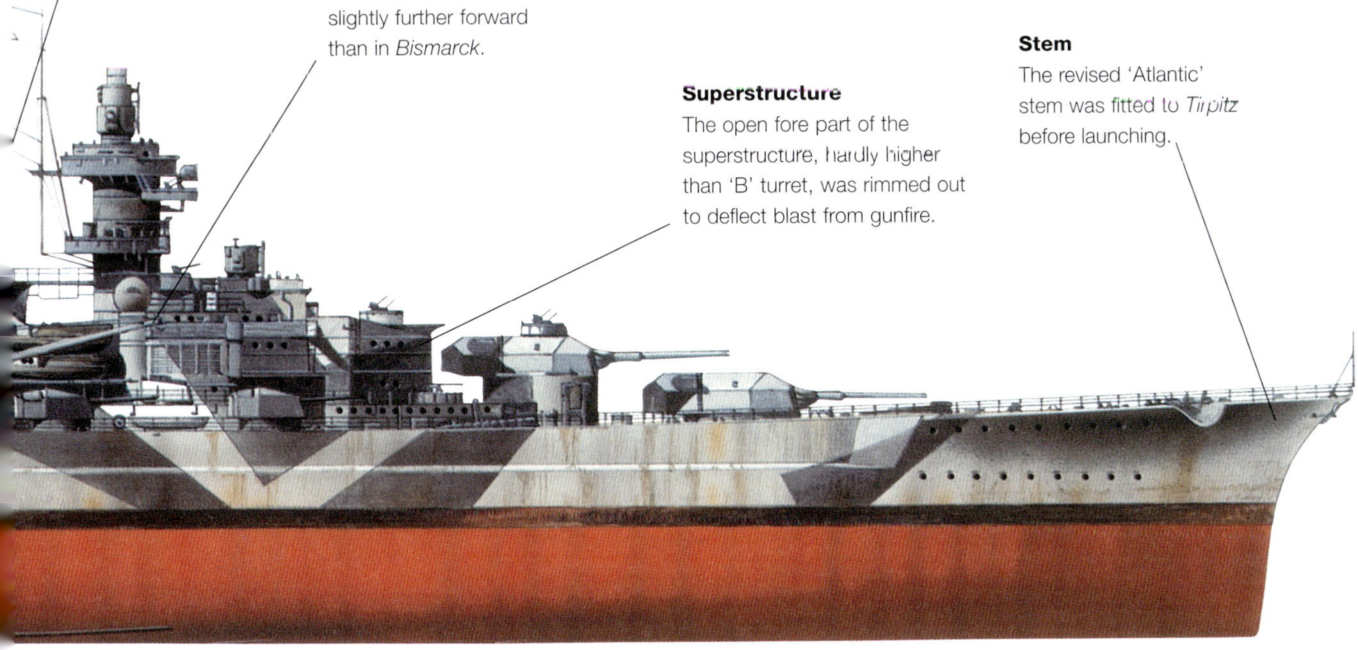

1943–44, a '*Würzburg*' radar set was installed on the after AA gunnery control position, with a 3m (9ft 9in) parabolic reflector for measuring height, and a Model 30 *Hohentwiel* radar aerial was mounted in the topmast.

Baltic Fleet

Tirpitz would have been just as formidable as *Bismarck*, but its potential career as a fighting ship was distorted by its symbolic importance. After successful sea trials it joined the Baltic Fleet and remained with it until January 1942, when it was moved to northern Norway as a guard against invasion and in order to attack Russia-bound convoys.

In March 1942 it fought off air attacks from the British carrier *Victorious*, shooting down two and was again unsuccessfully attacked from the air at Trondheim on 27–28 April. On 6–9 September 1942 its 380mm (15in) guns bombarded fortifications on Spitzbergen, but on the 22nd it sustained serious damage from attack by British midget submarines in Altafjord, near Narvik, and was out of action until March 1944. Russian as well as British aircraft made attacks, but the British were more successful, again causing serious damage on 3 April 1944 when 14 bomb hits were recorded.

The deck plan is essentially the same as that of *Bismarck*. Comparison with the cutaway profile below shows how much space was taken up by machinery, barbettes and magazines.

As long as *Tirpitz* remained in northern Norway there was a serious threat to Russian and Atlantic convoys, and British efforts to immobilise it continued, hampered by the extent of AA defences not only aboard the battleship but in the surrounding area. They achieved some success on 15 September 1944 when 4572kg (10,079lb) bombs, used for the first time, left the battleship afloat but unseaworthy. *Tirpitz* was moved under its own power to Tromsö. By now the only future that could be prepared for was to use the ship as a floating battery.

Operation Catechism

The attacks continued. On 12 November 1944, in 'Operation Catechism', 32 Lancaster bombers of the RAF's 9 and 617 Squadrons, from a base in Russia, converged

This cutaway view of *Tirpitz* shows the position of the eight 38mm (15in) L/52 guns arranged in four twin gun turrets: two superfiring turrets forward – 'Anton' and 'Bruno' – and two aft – 'Caesar' and 'Dora'.

Resilience

The fate of *Tirpitz* showed conclusively that the capital ship was no longer self-protectable; unless it could assume command of the air space around itself, it could only exist safely within a screen of anti-aircraft defences that itself negated the ship's offensive value. The Luftwaffe was severely criticised for failing to provide adequate air cover, but by 1944 Allied air superiority was firmly established. As did *Bismarck*, *Tirpitz* showed remarkable staying power under sustained attack. The 'Tallboy' weapons that finally sank it had not been envisaged at the time of its design, and had been devised by the 'dam-buster' scientist-inventor Sir Barnes Wallis specifically for the purpose of piercing its armour. Several hundred British and Russian aircraft made attacks over almost three years and many were shot down before the battleship was finally eliminated (see photo). *Tirpitz* did not have a 'glorious' career but its influence on the Arctic war was important.

on *Tirpitz* from 09:35. The ship's main guns opened up on them from long range, but there was no aerial opposition from the Luftwaffe. The Lancasters scored at least two direct hits and four near misses with 'Tallboy' 5080kg (11,199lb) bombs carrying around 3000kg (6614lb) of high explosive. The most destructive hit was between the aircraft catapult and funnel, which blew a massive hole in the side armour.

When *Tirpitz*'s armour was designed, it was envisaged that the maximum destructive force would be around 500kg (1100lb) of explosive. No imaginable battleship could survive such an assault. The force of the strikes was enough to burst open the hull side and cause the ship to capsize, and at the same time one of the main magazines exploded. Around 1000 men were lost.

Specification

Dimensions	Length 251m (823ft 6in), Beam 36m (118ft 1in), Draught 9.3m (30ft 6in), Displacement 38,283 tonnes (42,200 tons); 46,992 tonnes (51,800 tons) full load
Propulsion	12 Wagner superheated boilers, 3 Brown-Boveri geared turbines developing 102,906kW (138,000hp)
Armament	8 380mm (15in) guns, 12 150mm (5.9in) guns, 16 105mm (4.1in) guns, 16 37mm (1.5in) and 12 20mm (0.79in) AA guns
Armour	Belt 320–80mm (12.6–3.1in), Bulkheads 220–45mm (8.6–1.7in), Deck 120–80mm (4.3–1.7in), Barbettes 340–220mm (13.4–8.6in), Turrets 360–180mm (14.2-7in)
Range	16,430km (8870nm) at 19 knots
Speed	30 knots
Complement	2065

Indiana (1942)

The second ship of four in the *South Dakota* class, *Indiana* and its sister ships saw extensive action during World War II, all of it in Pacific Ocean campaigns.

A *South Dakota* class had been planned after World War I but cancelled under the Washington Treaty. This was a new design, incorporating numerous new features. *Indiana* (BB58) was laid down at Newport News on 20 November 1939, launched on 21 November 1941 and completed on 30 April 1942, at a cost of around $77 million. Completion was speeded up after Pearl Harbor, when new battleships were vitally needed.

Nine 406mm (16in) guns were carried in three triple turrets. These guns, Mark VI, had been designed in 1936 and were already installed on the preceding *North Carolina* and *Washington*. Each triple turret weighed 1549 tonnes (1708 tons). Calibre was L/50, the barrels weighed 86 tonnes (94.7 tons) and had a range of 38,700m (42,320yd) elevated to 45 degrees. *Indiana*'s secondary armament was 20 127mm (5in) guns in twin turrets, 24 40mm (1.6in) four-barrelled AA guns (a further eight were installed in 1943) and 50 20mm (0.79in) single-mount guns (reduced to 40 from 1943). With *North Carolina*, the US Navy ceased to instal torpedo tubes on its battleships.

The *South Dakota* class were the first US battleships to have a single funnel from the beginning, made possible by redesign of the engine space, which was 17m (55ft) shorter than in *North Carolina* though the engines were more

The *South Dakota* class was relatively compact in form – 210m (680ft) long compared to the preceding *North Carolina*'s 222.1m (728ft 8in) – but of almost the same beam, and capable of considerably greater speed.

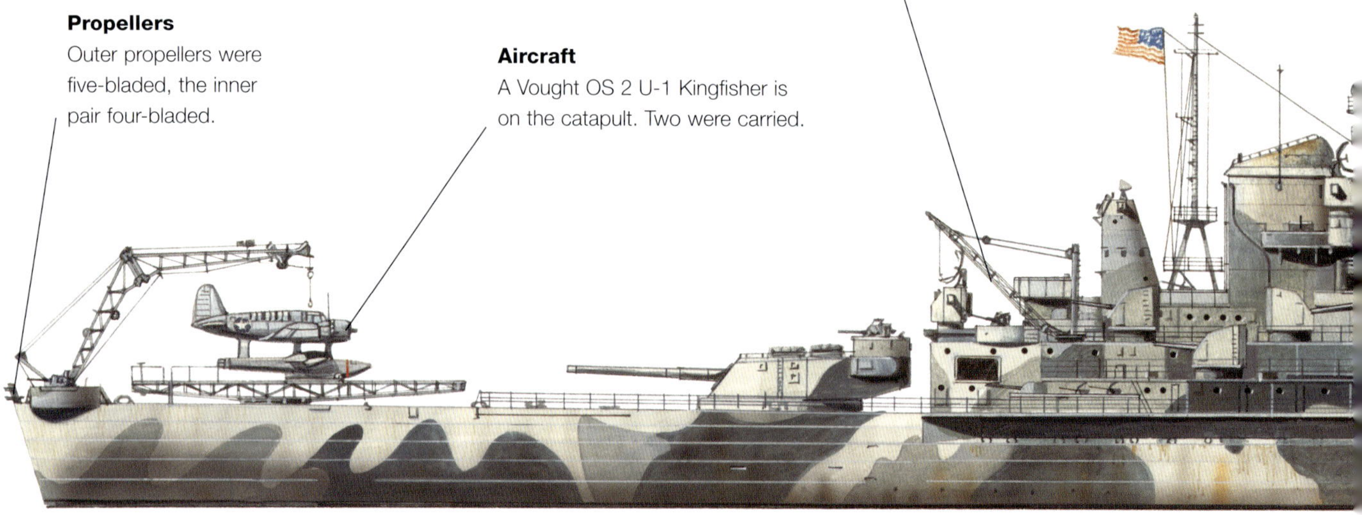

USS Indiana at anchor in Hampton Roads, Virginia, on 8 September 1942. Work appears to be in hand on the 406mm (16in) guns.

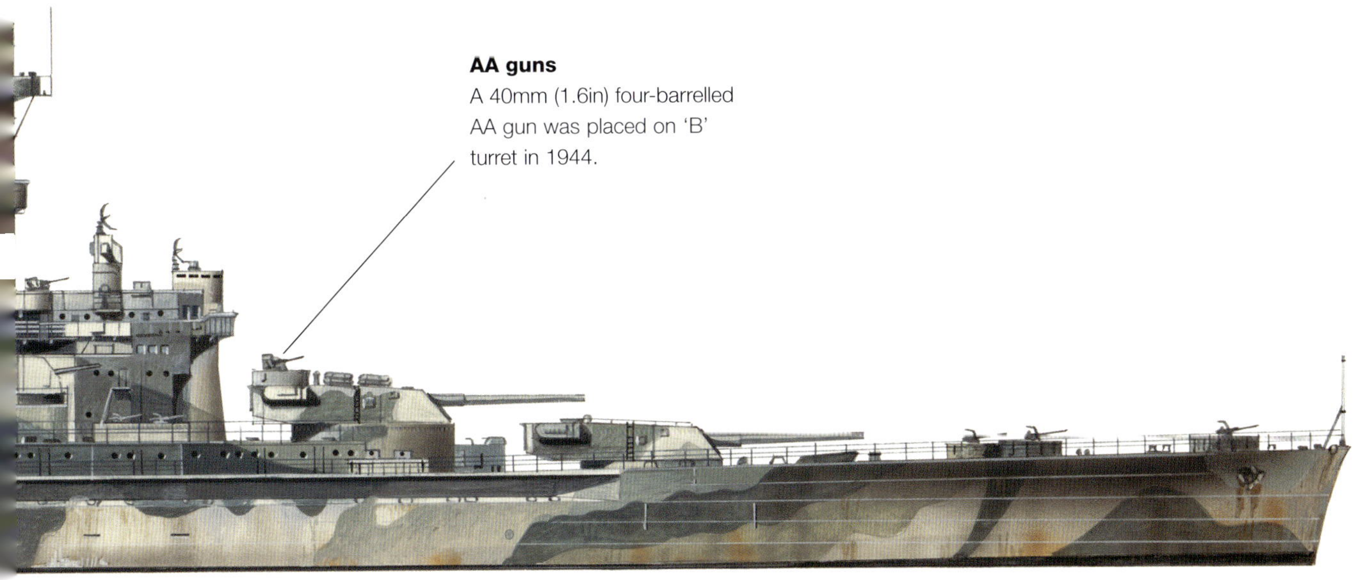

AA guns
A 40mm (1.6in) four-barrelled AA gun was placed on 'B' turret in 1944.

powerful. The ships were also shorter than their predecessors, though with the same beam, giving them a compact look that was accentuated by a higher freeboard and the mounting of the 127mm (5in) guns a deck above than the flush deck.

These were the first American ships to be given inclined internal side armour, reaching from the armoured deck to the inner bottom and 310mm (12.2in) thick, tapering to 25mm (1in). This layout gave the ships a long indented inward-angled stretch of the central hull, just below the flush deck, rather reminiscent of the old casemate structure. Torpedo bulges were not fitted to this class, but a splinter protection deck placed 80cm (2ft 7in) below the main armour deck was a new feature. The combination of bridge, tower, mast and funnel, though similar, varied on each ship and was changed on several occasions. Initially all carried a rotatory rangefinder on the conning tower, and a Type Sra radar antenna was mounted just abaft. The after mast, mounted on a tripod base, was heightened in 1945. By the end of the war *Indiana* was fitted with SG radar (aft), and SK-2 forward.

Guadalcanal campaign

Indiana's war service was all with the Pacific Fleet. From August 1942 it was involved in the fighting at Guadalcanal, and from 28 November it was engaged on carrier escort duties, which required relatively high speed. It supported the US landings on the Gilbert Islands and in January 1944 bombarded Kwajalein prior to the landings on the Marshall Islands. On the night of 1 February it collided with the battleship USS *Washington*, killing four men. Damage control parties went into action and some compartments were flooded to counteract a 4 degree list. Makeshift repairs

South Dakota's AA batteries in action against Japanese torpedo bombers in the Battle of the Santa Cruz Islands, 26 October 1942.

Ship building

The four ships of the class were all built within three years, a notably short time for capital ships. Co-operation between the four building yards, New York Shipbuilding, Newport News Shipbuilding, Bethlehem and Norfolk Navy Yard, helped, as did a sense of urgency about the rapidly deteriorating international situation even before the Pearl Harbor emergency. Two ships only were planned at first, but a further two were authorised on 25 June 1938. The original design had envisaged a speed of 41.7km/h (22.5 knots) but by 1939 the US Naval command had accepted that speed was also a requisite for a modern battleship (spurred on by reports of fast battleships under construction in Japan). This was fortunate, as so much of *Indiana*'s wartime duties consisted in escorting aircraft carriers at full speed.

were supplemented at Pearl Harbor. On 29–30 April *Indiana* was back on duty, joining Task Force 58 for the Truk Atoll raids, and on 19 June was hotly involved in the Battle of the Philippine Sea, essentially a carrier battle, in which it shot down several enemy aircraft. It then returned to escort and landing support, at one time remaining at sea for 64 consecutive days.

In October 1944 it went to Bremerton in Washington State for refit and was not involved in the Leyte Gulf battles. Back at Pearl Harbor in December it resumed its former duties and also bombarded shore targets, including Japanese mainland sites. In May 1945 it fought off kamikaze attacks off Okinawa.

Combat record

Indiana claimed the downing of 18 enemy aircraft. It participated in 10 shore bombardments and earned 12 battle stars for its various operations and engagements in the war years. It was part of the Allied fleet in Tokyo Bay on 5 September, and returned to the USA at San Francisco on 29 September, moving on to Puget Sound Navy Yard for overhaul. In September 1946 it was placed on the reserve list and was the last of the *South Dakota* class to be decommissioned in September 1947. It was stricken on 1 June 1962 and sold for scrapping on 6 September 1963.

Specification

Dimensions	Length 210m (680ft), Beam 32.9m (107ft 8in), Draught 8.9m (29ft 3in), Displacement 34,446 tonnes (37,970 tons); 40,387 tonnes (44,519 tons) full load
Propulsion	8 Foster Wheeler boilers, 4 Westinghouse geared turbines developing 96,941kW (130,000hp)
Armament	9 406mm (16in) guns, 20 127mm (5in) guns, 24 40mm (1.6in) and 50 20mm (0.79in) AA guns
Armour	Belt 310–22mm (12.2–0.87in), Bulkheads 279mm (11in), Barbettes 439–287mm (17.3–11.3in), Turrets 457–184mm (18–7.25in), Deck 152–146mm (6–5.75in), Tower 406–184mm (16–7.25in)
Range	27,750km (15,000nm) at 15 knots
Speed	27 knots
Complement	1793

Iowa (1943)

Lead-ship of the US Navy's last and largest class of battleships, pennant number BB61, *Iowa* was planned and built as a 'supership', to have a combination of armament, speed and armour superior to the best of any other nation.

The US Congress authorised the building of six large, fast battleships in 1938, but preliminary design work had already been going on for three years. In the event only four were constructed, *Iowa*, *Misssouri*, *New Jersey* and *Wisconsin*. Work began at the New York Navy Yard on 27 June 1940, it was launched on 27 August 1942 and commissioned on 22 February 1943 – a remarkably short period for the first capital ship of a new design. Building cost was slightly over $100 million.

The four *Iowa* class ships differed from the preceding *South Dakota* class in significant ways. Armour protection was external and kept flush with the hull to reduce drag. The relation of beam to length was smaller – maximum width was governed by the requirement that the ship should fit the Panama Canal locks. Almost 9072 tonnes (10,000 tons) heavier than *South Dakota*, these were the fastest battleships the world had seen, with a speed of 33 knots, enabling all four to form a fast division.

Funnel
Black paint on the funnel caps was a post-World War II alteration.

Aircraft
In World War II the aircraft were Vought OS2U Kingfishers (from 1941) and Curtiss SC Seahawks (from 1944).

Torpedo protection
Below the waterline tanks filled with oil or water lined the side outside the belt armour to detonate torpedoes and absorb the shock.

Instead of independently mounted masts the foremast was attached to an upper level of the tower, and the aftermast was mounted on the aft funnel. Originally both were of pole form, but the aftermast, heightened in 1945, was replaced by a cantilevered tripod in 1948. In 1958 a pole mast was added to operate a derrick, following the removal of the aircraft crane.

The profile shows USS *Missouri* as it appeared in 1944–45. It was on board *Missouri* that the Japanese surrender was formally signed on 2 September 1945, ending World War II.

Guns
The 406mm (16in) guns were Mark VII, of 50-calibre compared to *Indiana*'s 45-calibre. On 20 January 1989 one of them fired a shell to a distance of 43,300m (26.9 miles), believed to be a record.

Specification

Dimensions	Length 270.4m (887ft 2in), Beam 33.5m (108ft 3in), Draught 11.5m (38ft), Displacement 47,173 tonnes (52,000 tons); 52,118 tonnes (57,450 tons) full load
Propulsion	8 Babcock & Wilcox boilers, 4 GE steam turbines developing 158,088kW (212,000hp), 4 screws
Armament	9 406mm (16in) guns in 3 turrets, 20 127mm (5in) guns, 60 40mm (1.6in) 4-barrelled AA guns
Armour	Belt 310mm (12.2in), Barbettes 440–287mm (17.3–11.3in), Turret faces 500mm (19.7in), Deck 152mm (6in)
Aircraft	3, replaced by helicopters (1949) and UAVs (1984)
Range	23,960km (12,937nm) at 12 knots
Speed	33 knots
Complement	1921

Anti-aircraft defences

By the time *Iowa* entered service, the vulnerability of capital ships to air attack was very well known to the Allied forces. Consequently a massive battery of AA guns was fitted. A length 48.5m (159ft) greater than *South Dakota* allowed for more mountings, and the extra 8164 tonnes (9000 tons) provided a more substantial platform. In 1943 it carried 80 40mm (1.6in) Bofors guns and 49 20mm (0.79in) AA Oerlikon cannon mounted on both sides of the superstructure. Three reconnaissance floatplanes were carried, launched from two stern-mounted catapults. Also vital was the radar detection equipment. In 1943 SK and SRA aerials were fitted on the foremast and in 1945 type SK-2 was added, with type SC-2 on the aft mast. Further additions followed, with necessary alterations to the mast configurations.

Battleship Division 7

After training off Newfoundland from August to October 1943 in anticipation of an action against *Tirpitz* (which never happened), in November *Iowa* carried President Roosevelt across the Atlantic to Casablanca for the first Teheran conference with Churchill and Stalin. From 2 January 1944 it was in the Pacific, as flagship of Battleship Division 7, operating as escort to carrier groups and support for amphibious operations, seeing action at Guam, the Marshall Islands, in the Philippine Sea with Fast Carrier Task Force 58 and off Luzon and Formosa. In March it took some hits from Japanese coastal batteries on Mili Atoll, and a propeller shaft was damaged by Typhoon Cobra on 15 December. From 15 January to 19 March 1945 it underwent a refit at Hunters Point Yard, San Francisco, before returning to active service at Okinawa from April to June. During July and August its 406mm (16in) guns shelled positions on the Japanese mainland, and it arrived off Tokyo on 29 August,15 days after the Japanese ceased hostilities, to serve as Admiral Halsey's flagship for the formal surrender on 2 September.

After 1945 there was much debate about the future role of the *Iowa* class ships. Their speed in particular kept them viable while many other battleships were mothballed or scrapped. All were fitted with modernised bridges and other modifications were made, primarily to accommodate new radar equipment. *Iowa* was flagship of the Fifth Fleet in 1945–46. From 24 March 1949 to 25 August 1951 it was put in reserve, then with the outbreak of the Korean War returned to action, relieving and supporting army units off the coast of Korea and shelling shore positions.

Nuclear capability

In October 1952 it underwent a refit at Norfolk Navy Yard, following which it remained on the eastern side with the Atlantic Fleet. In 1954 the four ships of the class formed Battleship Division 2. *Iowa* also served from January–April 1955 as flagship of the Sixth Fleet in the Mediterranean. As the Cold War with Soviet Russia intensified from the mid-1950s, *Iowa* and *Wisconsin* were equipped with shells carrying nuclear warheads with an explosive force equal to that of the Hiroshima atom bomb.

Decommissioned on 24 February 1958, it remained on reserve at Philadelphia until 1984, when President Reagan's

enlarged Navy policy restored it to service on 28 April. At this time the now-obsolete AA armament was removed and new weapons fitted, including four MK 141 quad cell launchers for Harpoon anti-ship missiles, mounts for 32 BGM-109 Tomahawk missiles and four Phalanx CIWS Gatlings for anti-aircraft and anti-missile defence. Eight Unmanned Aerial Vehicles (UAVs) replaced the spotting helicopters which themselves had replaced floatplanes.

Original deck plan: comparison with the photo above reveals significant alterations to weaponry and aircraft installations between 1943 and 1952.

***Iowa* firing its full broadside to starboard against shore targets in North Korea, in mid-1952. The blast effect on the sea's surface is very noticeable.**

Iowa served in the Persian Gulf during the Iran–Iraq War of 1987, but its return to service was cut short by an explosion in No.2 turret on 19 April 1989, which killed 47 men. Finally decommissioned in 1990 and struck from the Naval Register in 1995, it was nominally reinstated between 1999 and 17 March 2006. It is now a museum ship at the Port of Los Angeles.

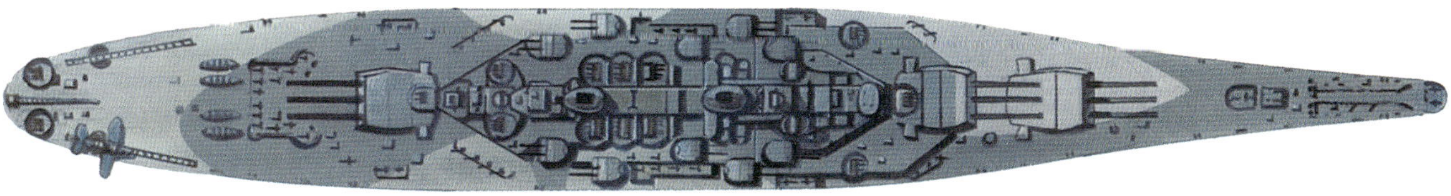

Vanguard (1946)

Great Britain's last and most powerful battleship was completed too late to be in action during World War II. Although a potent symbol of naval power, the changing requirements of the post-imperial Royal Navy left it essentially as a battleship without a role.

Orders for four battleships to form the *Lion* class, placed in 1938–39, had been suspended in 1939 and later cancelled. They were to have had new 406mm (16in) guns. *Vanguard* was a one-off, laid down at John Brown's yard in Clydebank on 2 October 1941, launched on 30 November 1944 and completed on 25 April 1946, at a cost of £11,530,503. Its 381mm (15in) guns, mountings and turrets were identical to those on HMS *Queen Elizabeth*, having been removed from HMS *Courageous* and *Glorious* on their conversion to aircraft carriers, and kept in store.

Modifications included heightening of the turret faces to allow for elevation to 40 degrees, and rangefinders were laid across the turret roofs. In addition 16 133mm (5.25in) guns were also carried in twin turrets, and 11 single-mount 40mm (1.6in) Bofors guns, one two-barrelled and 10 six-barrelled 40mm Bofor guns, all power-operated, with a maximum elevation of 90 degrees. The multiple Bofors guns were removed in 1954. Also carried were four 3-pounder guns for saluting purposes. No torpedo tubes or planes were carried.

Turrets
Vanguard's 381mm (15in) guns and turrets were of World War I construction.

Helicopter landing
Although no plane was carried, helicopters landed on the forecastle and after deck.

Lengthy construction

Much of the *Lion* and *King George V* class design work was incorporated, but during the long construction time numerous changes were made as a result of war experience. A straight bow and level forecastle, as in *King George V*, was replaced by a handsome cutaway bow with considerable flare, which helped to make *Vanguard* an exceptionally good sea-boat. It was also the only twentieth century British battleship with a flat transom stern. Armour and protective equipment were very much on the lines of the *George V* class, though internal subdivision and underwater protection were both much improved following the sinking of *Prince of Wales*.

The power plant was formed into four units, each of two boilers with a separate set of turbines, able to operate independently of the others, and the distance between inboard and outboard propeller shafts increased to 15.7m (51ft 6in) so that a single torpedo could not affect both. The armoured deck (as in the *King George V* ships) was 152mm (6in) thick over the magazines, capable of resisting the impact of a 450kg (1000lb) armour-piercing bomb coming from a height of 4300m (14,000ft); and 133mm (5.25in) over the machinery rooms, tapering to 63mm (2.5in). Angled caps were fitted to the two funnels from the beginning. Four 480kW (640hp) turbo generators and four 450kW (600hp) diesel generators supplied electric power

Vanguard's best aspect was the hull design, making it an exceptionally steady ship in rough seas, but in terms of armament and propulsion machinery it was less well equipped than the contemporary American _Iowa_ and Japanese _Yamato._

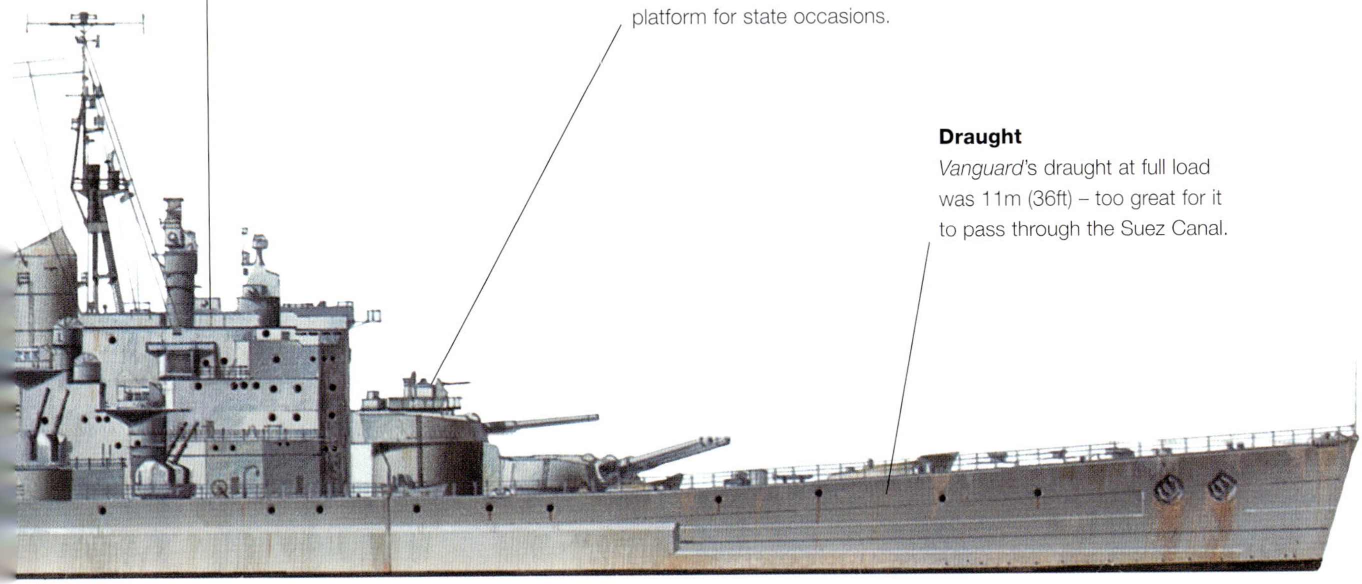

HMS *Vanguard* being broken up at Faslane, Scotland, 1960. She was decommissioned on 7 June 1960 and sold to the Iron and Steel Corporation of Great Britain for £560,000. By mid-1962 the last great battleship of the Royal Navy had ceased to exist.

Last battleship

A leading historian of warships wrote (after *Vanguard's* commissioning) that 'Nothing radically new had been created since the resumption of battleship construction in the 1930s... There was no denying that the end had been reached in their scope for development.' *Vanguard* was the last classic battleship to be built; the atomic era had already begun by the time it was commissioned. Its slow building history already reveals the decline in the importance of the battleship in war, though other factors contributed to the delay. Frigates, destroyers and submarines all were more urgently needed. If Germany had had more battleships, more vigorously deployed, *Vanguard* might have been completed as early as 1943; as it was, it did not join the fleet until a year after the end of the war, and Britain's biggest and best battleship was never put to the ultimate test.

at 220v, the most powerful electric installation of any British battleship. Air conditioning was installed in control rooms and some other areas, including the computing and transmission room located deep within the hull on the platform deck.

Radar

Several types of radar were fitted: AR262 (AA fire control), SW268 (surface warning), SR274 (main gun fire control), AR275 (secondary gun fire control), AW277 (height-finding), TI293 (target indication) and GW960 (air warning). This range of sensory and control equipment was unique in British battleships. There were two director control towers for the 406mm (15in) guns, each able to control all four

Specification

Dimensions	Length 249m (814ft 6in), Beam 32.8m (107ft 7in), Draught 9.3m (30ft 6in), Displacement 41,871 tonnes (46,155 tons); 46,665 tonnes (51,440 tons) full load
Propulsion	8 Admiralty 3-drum superheated boilers, 4 Parsons single reduction gear turbines developing 96,941kW (130,000hp)
Armament	8 381mm (15in) MkI guns, 16 133mm (5.25in) guns, 73 40mm (1.6in) Bofors AA guns
Armour	Belt 355–330mm (14–13in), Bulkheads 305–101mm (12–4in), Deck 152–133mm (6–5in), Barbettes 330–279mm (13–11in), Turrets 330–152mm (13–6in)
Range	16,670km (9000nm) at 20 knots
Speed	29.75 knots
Complement	1500

turrets. Four American-design director control towers controlled the 133mm (5.25in) guns, each with a twin-dome radar antenna.

Steam heating was provided in the turrets, look-out points and other exposed positions to enable the ship to operate effectively in Arctic waters. Amenities for the crew were superior to anything seen before: there was a chapel, a cinema, a dental surgery and schoolrooms.

Tugs cluster round *Vanguard* after the battleship grounded when leaving Portsmouth naval dockyard at the start of its last voyage, to the shipbreakers, on 4 August 1960.

Ceremonial duties

In the later 1940s much of *Vanguard*'s use was ceremonial, including a royal visit to South Africa in 1947. From March to July 1949 it was flagship of the Mediterranean Fleet, then of the Home Fleet Training Squadron from 12 November. On 13 September 1950 it was flagship of the Home Fleet, a role it alternated with that of flagship of the Royal Navy's Training Squadron.

While docking at Gibraltar on 10 February 1951 the carrier Indomitable collided with *Vanguard*, inflicting minor damage on the stern. During 1952–54 it participated in NATO exercises with the ships of the US, Dutch, French and Danish navies and made visits to ports in Norway and Sweden; it also played a prime role in the Coronation fleet review of 15 June 1953.

On 25 September 1954 it went to Devonport for a refit and preparation for reserve status, and from 28 November 1955 became flagship of the Reserve Fleet, making no further voyages. It was decommissioned on 7 June 1960 and towed to Faslane in Scotland for scrapping.

General Index

General Index

Ships Index

Ships Index